Cosmos-Universe-Particle-Gambol Theory

The Cosmos Theory Distribution of Universes
Anti-Planckian Model of Hubble Expansion Parameter
Our Universe as a Megaverse Particle
Universe-Megaverse Boundary
Implicit Evidence for a Megaverse
Casimir Force Limit on Hubble Expansion
S_8 Universe-Gambol Clumping Models
Variation of Universe Cluster Mass-Energy Sizes
Quantum Field Theory Derivation of Particle-Gambol Model
Quantum Field Theory Derivation of Megaverse-Gambol Universe Model
Quantum Field Theory Derivation of Cosmos-Universe-Gambol Model
The Cosmos Parent of Gambol Universes
Neutron Star Quark Core Gambol Model
GVDM Photon – Rho Gambol Model
Higgs and Massive Vector Boson Gambol Models

Stephen Blaha Ph. D.
Blaha Research

Pingree-Hill Publishing
MMXXIV

.

ISBN: 978-1-7356795-2-5

Rev. 00/00/01 January 3, 2024

To My Parents

Stephen and Wanda Blaha

Some Other Books by Stephen Blaha

SuperCivilizations: Civilizations as Superorganisms (McMann-Fisher Publishing, Auburn, NH, 2010)

All the Universe! Faster Than Light Tachyon Quark Starships & Particle Accelerators with the LHC as a Prototype Starship Drive Scientific Edition (Pingree-Hill Publishing, Auburn, NH, 2011).

Unification of God Theory and Unified SuperStandard Model THIRD EDITION (Pingree Hill Publishing, Auburn, NH, 2018).

The Exact QED Calculation of the Fine Structure Constant Implies ALL 4D Universes have the Same Physics/Life Prospects (Pingree Hill Publishing, Auburn, NH, 2019).

Integration of General Relativity and Quantum Theory: Octonion Cosmology, GiFT, Creation/Annihilation Spaces CASe, Reduction of Spaces to a Few Fermions and Symmetries in Fundamental Frames (Pingree Hill Publishing, Auburn, NH, 2021).

Passing Through Nature to Eternity ProtoCosmos, HyperCosmos, Unified SuperStandard Theory (Pingree Hill Publishing, Auburn, NH, 2022).

HyperCosmos Fractionation and Fundamental Reference Frame Based Unification: Particle Inner Space Basis of Parton and Dual Resonance Models (Pingree Hill Publishing, Auburn, NH, 2022).

The Cosmic Panorama: ProtoCosmos, HyperCosmos, Unified SuperStandard Theory (UST) Derivation (Pingree Hill Publishing, Auburn, NH, 2022).

A New Completely Geometric SU(8) Cosmos Theory; New PseudoFermion Fields; Fibonacci-like Dimension Arrays; Ramsey Number Approximation (Pingree Hill Publishing, Auburn, NH, 2023).

God and Cosmos Theory (Pingree Hill Publishing, Auburn, NH, 2023).

Newton's Apple is Now The Fermion (Pingree Hill Publishing, Auburn, NH, 2023).

Cosmos Theory: The Sub-Particle Gambol Model (Pingree Hill Publishing, Auburn, NH, 2023).

Available on Amazon.com, bn.com, Amazon.co.uk and other international web sites as well as at better bookstores.

CONTENTS

FIGURES and TABLES

Introduction

Previously, the author showed a gambol formalism from Cosmos Theory provided a detailed understanding of the form of the Deep Inelastic electron-proton scattering structure function, and also of quark, lepton, and hadron decay and scattering. This book provides a quantum field basis for that work. This book extends the theory of gambols in the study of particles AND universes.

The below points give gambols an aspect of reality. Quarks were accepted to explain hadrons although they are confined to hadrons and thus not independent particles in the everyday sense. Gambols are probabilistically confined at a deeper level within quarks, leptons and universes.

Gambols are more abstract since they exist as structure within fundamental particles (and composite particles). They seem to exist with highest probability as 8 gambol structures within fundamental particles AND as 1024 gambol structures within gamboled universes. They exhibit a new form of hidden Reality.

This book develops many new points:

1. A quantum theory of universes as particles described by a set of gambols.

2. The gambol theory gives a gambol model for the Hubble expansion of the universe based on a new anti-Planckian distribution. The expansion has an initial burst followed by a slow-growth interregnum, which then resumes a larger growth. It explains the "Hubble Tension."

3. A model describing the gamboled growth of mass-energies clumps within our universe. The model offers an understanding of S_8 clumping.

4. A model for universes within a Megaverse where the universes are gambols. Our universe then exists as a gambol. (Not a gamble!)

5. A model for universe expansion through the interplay of a Megaverse-based Casimir force and the force of internal universe matter-energy pressure.

6. A model for all the universes of all ten Cosmos spaces. Our universe is one of 5 sibling universes. The model probabilistically makes the number of Cosmos spaces to 10 using a Planckian distribution. There are 23 Cosmos universes in this model.

7. A quantum field theory for particles that embeds quantum gambols within particles as described in the author's Sub-Particles book. Quantum gambols replicate their parent particle's spin and internal symmetries.

8. A free gambol quantum field theory.

9. Gambols have a mass equal to the parent particle's mass divided by 8.

10. Universe gambols have a mass equal to the universe's mass divided by 1024.

11. Gambols have a temperature equal to a constant times the parent particle's (or parent universe's) mass. This feature is analogous to Black Hole temperature.

12. The quark core of a neutron star is described in a gambol model.

13. The relation of the pure electromagnetic photon and its rho meson attribute of GVDM are described in a gambol model

14. The acquision of mass by an ElectroWeak vector boson is described by a gambol model.

Highly recommended for serious readers.

1. Implications of the Gambol Models

The Gambol Model developed in Blaha (2023e) *Cosmos Theory: The Sub-Particle Gambol Model and* 2023d, *Newton's Apple is Now the Fermion* has numerous implications which we list below: The fractionation process of a particle into sets of gambols is described in Blaha (2023d) and (2023e).

1. Fundamental Fermions, Quark and Lepton particles, are composed of probabilistically determined sets of gambols that embody the interior and dynamics of the particle.

2. The probability of each set of gambols within a fermion of mass m is determined by a Planckian distribution:

$$U_{gi}(\varepsilon(s)) = 15\, N(\pi k T_g)^{-4}\, \varepsilon^3/(\exp(\varepsilon/(kT_g)) - 1)$$

   where

$$\varepsilon(s) = [(m_g s/m + 1)/(s + 1)]E$$
$$m_g = m/s$$

   is the gambol set energy ε(s) with E the individual gambol dynamic energy, and m_g the gambol mass. Each set of gambols labeled s, initially numbers $s = 2^n$ gambols. Each gambol of a set constitutes a universe of the Limos sector of the Cosmos spaces. ε(s) interpolates between the fermion mas and the gambol mass.

   $U_{gi}(\varepsilon(s))$ effectively gives the average number of gambols in a particle for each energy ε(s).

   In the case of universes (chapter 6 below) there is a modified ε(s) designed to have an infinite value at the point of the Big Bang (t = 0).

3. The label s is made continuous in calculations.

4. Gambols are confined to the interior of fermions and bosons.

5. Due to their probabilistic occurrence[1] within fermions, gambols are a new form of particle that only exist probabilistically.

---

[1] Gambols are thus a theoretical construct. We may view a fundamental particle as a probabilistic sum of fractionations into gambols—with each fractionation having a certain probability. In contrast a hadron is composed of a specific number of quarks that are "real" although they are confined within the hadron. Gambols are similarly confined but the gambol content of a particle varies with the fractionation—thus their "unreality." The gambol concept, which applies to all fundamental fermions, brings our understanding of the nature of matter to a deeper level. All particles are composed of gambols—a much "finer" substance of Reality. We see gambols → fundamental particles[1] → elementary particles → matter. The Gambol Model is supported by experimental evidence in Blaha (2023e).

6. The interior of a fermion has a gambol temperature that appears in the fermion's Planckian distribution. The temperature T_g satisfies the law

$$kT_g = 0.0785 \text{ m}$$

where m is the fermion's mass. The constant 0.0785 appears to be related to the confinement of gambols to within fermions.

In the case of universes (chapter 4 - 7 below) there is a modified kT_g designed to give kT_g a time dependent value.

7. The total set of gambols in a fermion may be viewed as a gas subject to the laws of Statistical Mechanics and Thermodynamics.

8. Each boson particle acts as a gambol made real.

9. The gambols of a hadron may be viewed as an individually combined composite of the gambols of its individual quarks or as combined into an overall composite for the hadron.

10. When a fundamental fermion or hadron interacts it may be viewed as undergoing a gambol interaction with the features:

    a. The interaction is a sum over sets of gambols probabilistically by individual gambols.
    b. Each gambol inherits all features of the parent particle except mass and momentum which are the gambol's mass and momentum.
    c. The interaction is the lowest order interaction without higher order terms for high energy interactions. We extrapolate this feature to lower energies.
    d. The S-matrix for the interaction has a Planckian probability factor for each gambol set $U_g(\varepsilon(s))$:

$$S_{fi} = \sum_{\substack{\text{Gambol} \\ \text{Sets}}} \prod_g U_g(\varepsilon(s)) S_{hadg}$$

where S_{hadg} is the gambol interaction S-matrix factor.

11. Within a fundamental fermion or hadron there is an internal gambol temperature $T_{g.}$ where

$$kT_{g.} = 0.785\text{m}$$

where m is the mass of the parent particle and k is Boltzmann's constant. The 0.785 factor is shown to originate in the gambol confinement mechanism described in Blaha (2023e).

12. When a resonance is created or decays the instantaneous gambol temperature of the initiating (or end) particle and the resonance are equal, as are the energies ε(s) with

$$kT_i = 0.0785\ m_{resonance}$$

13. When hadrons interact the gambols of each particle interact with those of other particles. The resulting S-matrix is a product of Planckian gambol probabilities and an interaction S-matrix element for interacting gambols of gambol masses and momentums. These terms are summed over the individual gambols.

14. Treating universes as decaying particles of enormous mass-energy, which are composed of gambols, we define a universe gambol temperature with

$$kT_{gu} = 0.0785 m_u\ s = 0.0785\ m_u/(bt)$$

where s = bt gives a steadily decreasing temperature with time that mirrors the decline of the standard universe temperature to T = 2.72548° K. (See chapter 6.) The constant $b = 8.66685 \times 10^{-19}\ sec^{-1}$.

15. We take the *total* mass-energy of the universe (including Dark mass and Dark energy) to be

$$m_u = 1.712 \times 10^{81}\ GeV/c^2$$

The above points summarize the gambol models of elementary particles presented in Blaha (2023e). In this volume we develop Gambol Models for universes with some slightly different features – due to physics of the Big Bang of universes and the structure of the set of universes of the HyperCosmos spaces.

2. Free Gambol Quantum Field Theory

The Limos based fractionation process may be applied to individual or sets of creation/annihilation operators, fermions, bosons, and dimensions. In this chapter we develop a free quantum field theory for fermion gambols. In chapter 6 of Blaha (2023e) we saw that deep inelastic electron-proton scattering data could be well represented by a sum over virtual photon – gambol scatterings. Thus it is relevant to develop a quantum field theory of gambols engaging in interactions.

Fermion gambols were shown to be confined in chapter 13 of Blaha (2023e) due to the Casimir effect of the Fermion Dirac vacuum. This argument cannot be maintained for boson gambols since bosons do not have a Dirac sea. Thus we provisionally view bosons as each consisting of a single gambol.

2.1 Fermion Gambol Planckian Distribution

Gambols were shown to have an associated Planckian probability distribution in Blaha (2023e). This was verified in the case of e-p deep inelastic scattering by showing the structure function F(ω) data was well fit by the gambol Planckian distribution.[2] The gambol Planckian distribution was also shown to fit quark, lepton and hadron scattering and decay.

Having shown the successful role of the gambol Planckian distribution we now consider its form in more detail before using it to develop a particle-gambol quantum field theory.

The particle[3] Planckian distribution $U_g(\varepsilon(s, p))$ is specified by

$$\begin{aligned} U_g(\varepsilon(s, p)) &= 15\,N\,(\pi\,kT_g)^{-4}\,\varepsilon^3/(\exp(\varepsilon/kT_g) - 1) \\ \varepsilon(s, p) &= [(m_g s/m + 1)/(s + 1)]p^0 \\ m_g &= m/s \\ kT_{g.} &= 0.0785\text{ m} \end{aligned} \qquad (2.1)$$

where a fermion is fractionated into s gambols. For each value of s there is a set of s gambols. A particle is a distribution of sets of gambols of varying fractionation s. Each gambol in a set has identical features. When a particle interacts, one, and only one, of its gambols participates individually in the interaction as the representative of the parent particle. The gambol that interacts has the mass $m_g = m/s$ and the momentum $p_g = p/s$.

The constant m is the mass of the particle within which the gambol resides, T_g is the gambol temperature,[4] and k is the Boltzmann constant.

---

[2] Since the experimental data does not lie on a line but rather has a spread for each value of ω the fit must be viewed as an approximation. When data that lies on a specific line appears, the fit may be simply adjusted to the new data. The appearance of a Q^2 dependence beyond scaling in x would require a further adjustment in the gambol distribution.

[3] In the case of universes, we use an anti-Planckian distribution that will be derived in chapter 6 since the time variation of the changes of a universe are different from the "instantaneous" changes of state for particles.

The distribution gives a probability[5] to the sets of gambols obtained by fractionating a fermion particle into gambols. The fractionation is dependent on the overall nature of the particle's interactions. In the case of deep inelastic scattering we found

$$s = 2M\omega \tag{2.2}$$

where

$$\omega = \nu/Q^2$$

Thus the fractionation was dependent on the energy momentum factor ω.

In the case of quarks, leptons, and hadron interactions $s = m/m_g$ is the measure of the fractionation. Thus as s increases one probes more deeply into the particle to decreasing gambol masses.

In the case of Hubble universe expansion (chapter 6) the fractionation s dependent on the time (eq. 6.10):

$$s = 1/(bt) \tag{2.3}$$

Time and energy are complementary variables. We find increasing s gives a deeper probe into particles and universes. The depth is marked by the stages of sets of gambols.

The probability $U_g(\varepsilon(s, p))$ gives the dependence of particle and interaction features on the depth (fractionation) as one progresses through increasing s values.

The gambol energy $\varepsilon(s, p)$ of the sth fractionation

$$\varepsilon(s, p) = [(m_g s/m + 1)/(s + 1)]p^0 \tag{2.4}$$

is based on an interpolation from the gambol mass m_g and the particle mass m. It follows from the interpolating expression[6]

$$\varepsilon(s, p) = [(m_g s + m)/(s + 1)]/\surd(1 - v^2) \tag{2.5}$$

with c = 1. Note $m/\surd(1 - v^2) = p^0$. For s = 0,

$$\varepsilon(0, p) = p^0 = m/\surd(1 - v^2)$$

For large s,

$$\varepsilon(s, p) = (m_g/m)\, p^0 = m_g/\surd(1 - v^2)$$

Note:

$$m_g \to 0 \text{ as } s \to \infty \qquad \text{since } m_g = m/s$$

---

[4] In chapter 6 we use the known decline of temperature with time as the cause of the introduction of time in the gambol temperature $kT_{gu.} = 0.0785$ m/bt.

[5] The probability may be viewed as the count of the number of gambols of each energy ε in the particle.

[6] In the case of e-p deep inelastic scattering we assumed the gambol was at rest in the proton's rest frame. If it was moving, a Q^2 dependence would be introduced as is observed in recent e-p and n-p experiments.

For fundamental particles and hadronic interactions we set $m_g = m/8$ treating s = 8 as an accurate approximation to m_g. For the Hubble parameter calculation in chapter 6 we set $m_g = m/1024$ since it reflects the wider scope of universes.

The Planck distribution for black bodies described the distribution of massless photons of energy $\varepsilon = h\nu$. The Gambol Planckian distributions describe the probability (fractionation) distribution of gambols within a massive particle (universe) in terms of the interpolated/fractionated energy of gambols $\varepsilon(s, p)$. The distribution may be viewed as a count of the number of gambols for each energy $\varepsilon(s)$.

2.2 Fermion Particle-Gambol Quantum Fields

The goals of formulating a particle-gambol theory are

1. To provide a formalism for both the particle and possible sets of gambols.
2. To have the same particle and gambol spin and internal symmetries.
3. To have gambol masses and momenta different from those of the particle.
4. To provide a Gambol Planckian distribution for sets of gambols such as eq. 2.1.

These objectives are met in chapter 3.

2.3 Gambol Creation/Annihilation Operators

The use of sets of gambols parameterized by a fractionation s having a probability distribution for their appearance in fermions leads to a need for a new quantum Field Theory formulation.

We begin by considering the creation and annihilation operators of a gambol quantum field. Gambol states must have spin ½ gambols, a mass $m_g = m/s$ where m is an input mass that becomes the mass of the parent particle in chapter 3; and a momentum $p_g = m_g/m\ p$ where p will become the mass of the parent particle. (Chapter 3)

We define the PseudoFermion creation and annihilation operators:

$$b^{1/n}{}_{g\gamma i}(q, S) \text{ and } b^{1/n}{}_{g\gamma i}{}^{\dagger}(q, S) \tag{2.6}$$

and

$$d^{1/n}{}_{g\gamma i}(q, S) \text{ and } d^{1/n}{}_{g\gamma i}{}^{\dagger}(q, S)$$

where S is the spin, γ is the internal symmetry index, the integer i is the PseudoFermion index, and n is the value of the fractionation. The operators satisfy

$$\{b^{1/n}{}_{g\gamma i}(q, S), b^{1/m}{}_{g\gamma' j}{}^{\dagger}(q', S')\} = (1 - \delta_{ij})\, \delta_{S,S'}\, \delta_{mn} \delta_{\gamma\gamma'} \delta^{r-1}(\mathbf{q} - \mathbf{q'})\, U_g(\varepsilon(n,q)) \tag{2.7}$$

and

$$\{d^{1/n}{}_{g\gamma i}(q, S), d^{1/m}{}_{g\gamma' j}{}^{\dagger}(q', S')\} = (1 - \delta_{ij})\, \delta_{S,S'}\, \delta_{mn} \delta_{\gamma\gamma'} \delta^{r-1}(\mathbf{q} - \mathbf{q'})\, U_g(\varepsilon(n,q)) \tag{2.8}$$

for a universe with space-time dimensions r and spatial dimension r – 1 with U_g defined by eq. 2.1 and in section 2.4.

We interpret the creation operator $b^{1/n}{}_{g\beta i}{}^{\dagger}(q', S')$ as creating a set of n identical gambols of. the same mass, momentum, spin and internal symmetries. The operator projects one representative gambol to participate in interactions. Since it is a projection operator it satisfies the projection conditions:

$$
\begin{aligned}
b^{1/n}{}_{g\gamma i}{}^{\dagger}(q, S)^2 &= b^{1/n}{}_{g\gamma i}{}^{\dagger}(q, S) \\
d^{1/n}{}_{g\gamma i}{}^{\dagger}(q, S)^2 &= d^{1/n}{}_{g\gamma i}{}^{\dagger}(q, S) \\
b^{1/n}{}_{g\gamma i}(q, S)^2 &= b^{1/n}{}_{g\gamma i}(q, S) \\
d^{1/n}{}_{g\gamma i}(q, S)^2 &= d^{1/n}{}_{g\gamma i}(q, S)
\end{aligned}
\tag{2.9}
$$

There is no problem with the Pauli Exclusion Principle since multiples of the creation operator become a single instance. The use of projections is necessitated by

1. The need to limit the distribution for a particle's gambols set. A particle may only have one gambol distribution.

2. The requirement is to have only one gambol emerge to enter into interactions. The other gambols of a set are spectator gambols. They are excluded from interacting. The generated gambol has a fraction of a parent's momentum and mass. The silent partner gambols "flow" through an interaction unchanged. See Fig. 2.1 for a gambol deep inelastic example.

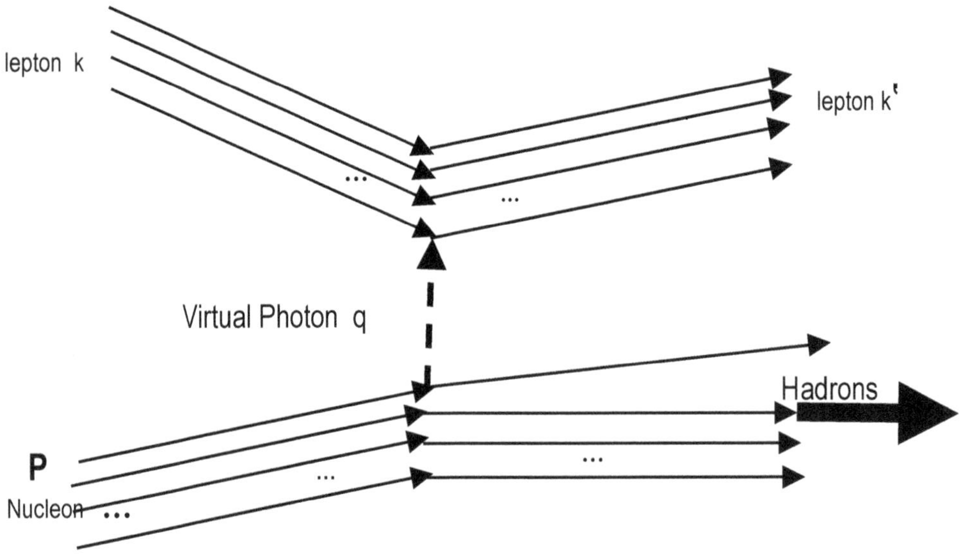

Figure 2.1. The kinematics of lepton-nucleon Deep Inelastic Scattering. The leptons, nucleons and virtual photon are all composed of partons of the Nth state. One gambol in the electron exchanges a photon with one gambol in the

proton. The other gambols are spectators. They contribute as part of the summation of these single gambol interactions.

The operator $b^{1/n}{}_{g\gamma 2}{}^{\dagger}$ creates a one particle state of one gambol within the set of n gambols:[7]

$$|1> = b^{1/n}{}_{g\gamma 2}{}^{\dagger}(q, S)|0> \qquad (2.10)$$

$$b^{1/n}{}_{g\gamma 2}{}^{\dagger}(q, S)^2\,|0> = b^{1/n}{}_{g\gamma 2}{}^{\dagger}(q, S)|0> = |1> \qquad (2.11)$$

with similar expressions for $d^{1/n}{}_{g\gamma 2}{}^{\dagger}(q, S)$:

$b^{1/n}{}_{g\gamma 2}(q, S)$ satisfies:

$$b^{1/n}{}_{g\gamma 2}(q, S)|0> = 0 \qquad (2.12)$$

and

$$N_n|k> = b^{1/n}{}_{g\gamma 2}{}^{\dagger}(q, S)b^{1/n}{}_{g\gamma 1}(q, S)|1> = |1> \qquad (2.13)$$

where N_n is the fermion number operator. Note N_n uses type 1 PseudoFermion operators.

When inner products are evaluated the probability distribution appears:

$$A_{nqS\gamma} = <n, q, S, \gamma|n', q', S', \gamma> = \delta_{n,n'}\delta_{S,S'}\delta_{\gamma\gamma'}\delta^{r-1}(\mathbf{q} - \mathbf{q'})\, U_g(\varepsilon(s,q)) \qquad (2.14)$$

$A_{nqS\gamma}$ is the amplitude for the one gambol state extracted from the set of n gambols. Note: U is normalized:

$$1 = \int_0^\infty d\varepsilon\, U(\varepsilon) \qquad (2.15)$$

Then we see the integrated $A_{nqS\gamma}$ is a sum over all possible gambol sets:

$$\int_0^\infty d\varepsilon\, A_{nqS\gamma} = \delta_{n,n'}\delta_{S,S'}\delta_{\gamma\gamma'}\delta^{r-1}(\mathbf{q} - \mathbf{q'})$$

The role of the creation operator is to create a fractionated set of gambol states. The state embodies all features except mass and momenta, which are fractional.

A two gambol state is

$$|2> = b^{1/n}{}_{g\gamma 2}{}^{\dagger}(q, S)\, b^{1/n}{}_{g\gamma 2}{}^{\dagger}(q', S')|0>$$

Each operator has different index numbers thus avoiding the projection relation of eq. 2.9.

2.4 Normalizing the Gambol Planckian Distribution

The black body Planck distribution for photons has the form:

[7] We routinely use the number 2 creation operator to create states in PseudoQuantum Theory.

$$u(\varepsilon) \equiv u(\nu) = 8\pi\varepsilon^3/h^2c^3/(e^{\varepsilon/kT} - 1) \qquad (2.16)$$
$$= 8\pi h\nu^3/c^3\ 1/(e^{h\nu/kT} - 1)$$

where ν is the frequency and

$$\varepsilon = \text{fractionation energy} = h\nu \qquad (2.17)$$

with h being Planck/s constant.

We normalize the integral of the Planck probability distribution to one:

$$1 = \int_0^\infty d\varepsilon\ c/(4\sigma T^4)\ u(\nu) \qquad (2.18)$$

where

$$\sigma = 2\pi^5\ k^4/(15h^2c^2) \qquad (2.19)$$

We define the *normalized* Planck distribution:

$$U(\varepsilon) = 15/(\pi kT)^4\ \varepsilon^3/(e^{\varepsilon/kT} - 1) \qquad (2.20)$$

with normalization

$$1 = \int_0^\infty d\varepsilon\ U(\varepsilon) \qquad (2.21)$$

resulting in the gambol *Planckian* distribution *for particles*:

$$U_g(\varepsilon(s)) = 15\ N\ (\pi kT)^{-4}\ \varepsilon^3/(e^{\varepsilon/kT} - 1) \qquad (2.22)$$

where N is an additional normalization constant. Inverting ε in eq. 2.4 we find s as a function of ε:

$$s(\varepsilon) = (1 - \varepsilon/p^0)/(\varepsilon/p^0 - m_g/m) \qquad (2.23)$$

Eq. 2.23 assumes $m_g \neq m/s$. In the study of particle Planckian distributions in Blaha (2023e) we found that the specific effective value[8] of s is 8 resulting in

$$m_g = m/8 \qquad (2.24)$$

We used that value in the many calculations in Blaha (2023e).

2.5 Gambol Quantum Fields

We next define fractional PseudoQuantum free fermion gambol fields in the space of Blaha number N:

---

[8] Later, when we apply a gambol distribution to universes (chapters 4 – 7) we use $m_g = m/1024$ as the universe gambol mass m_g.

$$\psi_{\gamma 1}{}^{1/n}(x) = \Sigma_\alpha\, [b^{1/n}{}_{g\gamma 1}\, f_\alpha(x) + d^{1/n}{}_{g\gamma 1}{}^{\dagger} f_\alpha{}^*(x)] \tag{2.25}$$
$$\psi_{\gamma 2}{}^{1/n}(x) = \Sigma_\alpha\, [b^{1/n}{}_{g\gamma 2} f_\alpha(x) + d^{1/n}{}_{g\gamma 2}{}^{\dagger}\, f_\alpha{}^*(x)]$$

where the $f_\alpha(x)$ are Fourier fields with α being the momentum in Blaha space N.

The fermion gambol Green's function $<0|T(\psi_1{}^{1/n}(x)\,,\, \psi_1{}^{1/n}(y)\,)|0>$ has U(ε(s,p)) within it. Its Fourier transform is

$$G(p) = U_{gi}(\varepsilon(s,\, p))/\gamma\cdot p \tag{2.26}$$

We describe the Lagrangian formulation of particles containing gambols in chapter 3. A particle-gambol joint formulation provides the correct context for gambols. Fermion gambols are confined to fermion particles as shown in Blaha (2023e).

Boson particles have boson gambols. In Blaha (2023e) we showed that boson gambols are not Dirac sea confined since there is no Dirac sea for gambols. We thus provisionally treat boson particles as consisting of a single gambol. The black body Planck distribution is a gambol boson distribution for massless gambols – photons. The Planckian gambol distributions are appropriate for massive particles and gambols.

3. Free Gambol Quantum Field Theory of Particles Containing Gambols

The previous chapter described quantized fermion gambol sets. This chapter describes a quantum field theory of fermions containing sets of gambols. When a fermion particle interacts, each of its sets of gambols generates a representative gambol to undergo the interaction. The "selected" gambol inherits all the features (spin and internal symmetry) of the parent particle except mass and momentum. The mass is a fraction m/s of the particle mass m where 1/s is the fractionation. Similarly the gambol momentum is p^μ/s where p^μ is the particle momentum.

3.1 PseudoFermion Quantum Theory

Following Blaha (2023c) we define[9] two coordinates PseudoFermions with the form

$$\psi_{i\alpha\beta}(y, z) \tag{3.1}$$

where y and z are independent coordinates in r space-time dimensions, where i = 1, 2 labels the PseudoQuantum[10] fields, and α and β are *spinor* indices.

We begin by defining a free PseudoFermion PseudoQuantum Lagrangian with two PseudoQuantum wave functions ψ_1 and ψ_2 that are functions of two sets of coordinates in r space-time dimensions, y and z,

$$\begin{aligned}\mathscr{L} = \overline{\psi}_{2\alpha\beta}[-M^{-1}\gamma_{y\alpha\kappa}{}^{\mu}\cdot\partial/\partial y^{\mu}\ \gamma_{z\beta\lambda}{}^{\nu}\partial/\partial z^{\nu} - M]\psi_{1\kappa\lambda} + \\ + \overline{\psi}_{1\alpha\beta}[-M^{-1}\gamma_{y\alpha\kappa}{}^{\mu}\cdot\partial/\partial y^{\mu}\ \gamma_{z\beta\lambda}{}^{\nu}\partial/\partial z^{\nu} - M]\psi_{2\kappa\lambda}\end{aligned} \tag{3.2}$$

where $\gamma_y{}^\mu$ and $\gamma_z{}^\mu$ are Dirac matrices for y and z coordinates respectively, α, β, κ, and λ are spinor indices, and y and z are coordinates in r = 4 dimension space-time, M is the mass, and

$$\overline{\psi}_{i\alpha\beta} = \psi_{i\kappa\lambda}{}^{\dagger}\ \gamma_y{}^0{}_{\kappa\alpha}\ \gamma_z{}^0{}_{\lambda\beta} \tag{3.2a}$$

---

[9] We follow the conventions of Bjorken (1965) with the $g^{\mu\nu}$ metric (1, -1, -1, -1).

[10] Appendix C of Blaha (2016f) presents a first quantized PseudoQuantum theory CQ Mechanics that embodies both classical and quantum theory. **This theory is the non-relativistic quantum mechanics limit of the relativistic PseudoFermion theory.** CQ Mechanics has two sets of coordinates that combine to create a generalization of conventional Quantum Mechanics. It has applications in a generalized Feynman path integral formalism, a generalized Schrödinger equation, a generalized Boltzmann equation, the Fokker-Planck equation, a generalized approach to quantum and classical chaos, and to quantum entanglement as well as semi-quantum entanglement. Our "Pseudo" formalisms apply to both Quantum Field Theory and Quantum Mechanics. In these applications there is a clear almost continuous transition between the quantum to the classical sectors.

for i = 1, 2 where the Dirac matrix subscripts y and z indicate they are associated with the y and z coordinates.

The equations of motion are

$$[-M^{-1}\gamma_y{}^{\mu}\cdot\partial/\partial y^{\mu}\,\gamma_z{}^{\nu}\partial/\partial z^{\nu} - M]\psi_1 = 0 \tag{3.2b}$$
$$[-M^{-1}\gamma_y{}^{\mu}\cdot\partial/\partial y^{\mu}\,\gamma_z{}^{\nu}\partial/\partial z^{\nu} - M]\psi_2 = 0 \tag{3.2c}$$

We define subsidiary equations of motion

$$[i\gamma_y{}^{\mu}\cdot\partial/\partial y^{\mu} - M]\psi_j = 0 \tag{3.2d}$$
$$[i\gamma_z{}^{\nu}\partial/\partial z^{\nu} - M]\psi_j = 0 \tag{3.2e}$$

for j = 1, 2. Eqs. 3.2d and 3.2e imply eqs. 3.2b and 3.2c.

One conjugate momentum is (with two spinor indices α, β)

$$\pi_{y1\alpha\beta} = \partial\mathscr{L}/\partial(\partial\psi_{1\alpha\beta}/\partial y^0) = -M^{-1}(\gamma_z{}^{\nu}\partial/\partial z^{\nu}\,\gamma_z{}^{0}\psi_2{}^{\dagger})_{\alpha\beta} = (\psi_2{}^{\dagger}\gamma_z{}^{0})_{\alpha\beta} \tag{3.3}$$

after partial integrations (with surface terms having the value zero) of

$$L = \int d^r y \int d^r z\, \mathscr{L}$$

using the subsidiary equation of motion:

$$\gamma_z{}^{\nu}\partial/\partial z^{\nu}\,\psi_2{}^{\dagger} = -M\psi_2{}^{\dagger} \tag{3.4}$$

Similarly the other conjugate momenta are

$$\pi_{z1\alpha\beta} = \partial\mathscr{L}/\partial(\partial\psi_{1\alpha\beta}/\partial z^0) = -M^{-1}(\gamma_y{}^{\nu}\partial/\partial y^{\nu}\,\gamma_y{}^{0}\psi_2{}^{\dagger})_{\alpha\beta} = (\psi_2{}^{\dagger}\gamma_y{}^{0})_{\alpha\beta} \tag{3.5}$$

and

$$\pi_{y2\alpha\beta} = \partial\mathscr{L}/\partial(\partial\psi_{2\alpha\beta}/\partial y^0) = -M^{-1}(\gamma_z{}^{\nu}\partial/\partial z^{\nu}\,\psi_1{}^{\dagger})_{\alpha\beta} = (\psi_1{}^{\dagger}\gamma_z{}^{0})_{\alpha\beta} \tag{3.6}$$
$$\pi_{z2\alpha\beta} = \partial\mathscr{L}/\partial(\partial\psi_{2\alpha\beta}/\partial z^0) = -M^{-1}(\gamma_z{}^{\nu}\partial/\partial y^{\nu}\,\psi_1{}^{\dagger})_{\alpha\beta} = (\psi_1{}^{\dagger}\gamma_y{}^{0})_{\alpha\beta} \tag{3.7}$$

using the subsidiary equations of motion:

$$\gamma_z{}^{\nu}\partial/\partial z^{\nu}\,\psi_1{}^{\dagger} = -M\psi_1{}^{\dagger} \tag{3.8}$$
$$\gamma_y{}^{\nu}\partial/\partial y^{\nu}\,\psi_1{}^{\dagger} = -M\psi_1{}^{\dagger} \tag{3.9}$$
$$\gamma_y{}^{\nu}\partial/\partial y^{\nu}\,\psi_1{}^{\dagger} = -M\psi_1{}^{\dagger} \tag{3.10}$$

We define new momenta to preserve y – z symmetry:

$$\pi_{1\alpha\beta} = (\pi_{y1}\gamma_z{}^{0})_{\alpha\beta} = (\pi_{z1}\gamma_y{}^{0})_{\alpha\beta} = \psi_2{}^{\dagger}{}_{\alpha\beta} \tag{3.11}$$
$$\pi_{2\alpha\beta} = (\pi_{y2}\gamma_z{}^{0})_{\alpha\beta} = (\pi_{z2}\gamma_y{}^{0})_{\alpha\beta} = \psi_1{}^{\dagger}{}_{\alpha\beta} \tag{3.12}$$

The form of the conjugate momenta implies the only non-zero equal time anti-commutators are:[11]

$$\{\pi_{j\alpha\beta}(\mathbf{y}, y^0, \mathbf{z}, z^0),\ \psi_{i\kappa\lambda}(\mathbf{y}', y^0, \mathbf{z}', z^0)\} = \{\psi_{j\alpha\beta}^{\dagger}(\mathbf{y}, y^0, \mathbf{z}, z^0),\ \psi_{i\kappa\lambda}(\mathbf{y}', y^0, \mathbf{z}', z^0)\}$$

$$= (1 - \delta_{ij})\,\delta_{\alpha\kappa}\,{}_{\beta}\delta_{\beta\lambda}\,\delta^{r-1}(\mathbf{y} - \mathbf{y}')\delta^{r-1}(z - \mathbf{z}') \qquad (3.13)$$

for i, j = 1, 2.

If we introduce fractionation s where the particle is fractionated to s gambols, and internal symmetry index γ, the free PseudoFermion wave function has the form:

$$\psi_{i\alpha\beta\gamma}(y, z) = \sum_{s}\sum_{s_1}\sum_{s_2} \int dp^{r-1}\int dq^{r-1}\, N(p, q)[b_{\gamma i}(s, p, q, s_1, s_2)u_\alpha(p,s_1)u_\beta(q,s_2)\exp(-ip\cdot y - iq\cdot z) +$$

$$+ d_{\gamma i}(s, p, q, s_1, s_2)^{\dagger}v_\alpha(p,s_1)v_\beta(q,s_2)\exp(ip\cdot y + iq\cdot z)] \qquad (3.14)$$

plus Hermitean conjugates for i = 1, 2 where s_i is the spin and N(p, q) is a normalization factor.

The composite creation and annihilation operators satisfy the anti-commutation relations:

$$\{b_{\gamma i}(s, p, q, s_1, s_2), b_{\gamma j}(s', p', q', s_1', s_2')^{\dagger}\} = \{d_{\gamma i}(s, p, q, s_1, s_2), d_{\gamma j}(s', p', q', s_1', s_2')^{\dagger}\}=$$

$$= (1 - \delta_{ij})\,\delta_{s,s'}\,\delta_{s_1,s_1'}\,\delta_{s_2,s_2'}\,\delta^{r-1}(\mathbf{p} - \mathbf{p}')\delta^{r-1}(\mathbf{q} - \mathbf{q}')\,U_g(\varepsilon(s,q)) \qquad (3.15)$$

where U_g is given by eqs. 2. 22.

The other composite anti-commutation operators are zeroes.

$$\begin{aligned}
&\{b_{\gamma i}(s, p, q, s_1, s_2), b_{\gamma j}(s', p', q', s_1', s_2')\} = 0 \qquad (3.16)\\
&\{b_{\gamma i}(s, p, q, s_1, s_2)^{\dagger}, b_{\gamma j}(s', p', q', s_1', s_2')^{\dagger}\} = 0\\
&\{d_{\gamma i}(s, p, q, s_1, s_2), d_{\gamma j}(s', p', q', s_1', s_2')\} = 0\\
&\{d_{\gamma i}(s, p, q, s_1, s_2)^{\dagger}, d_{\gamma j}(s', p', q', s_1', s_2')^{\dagger}\} = 0
\end{aligned}$$

and the anti-commutators of b and d type operators are zero as well.

3.2.1 Extraction of Gambol and Particle Operators

The creation/annihilation operators may be extracted from the wave function of eq. 3.14 by integrating over (r – 1) – coordinates:

$$\int dz^{r-1}e^{ik\cdot z}\psi_{i\alpha\beta\gamma}(y,z) = (2\pi)^{r-1}\{\sum_{s}\sum_{s_1}\sum_{s_2} \int dp^{r-1}N(p,k)[b_{\gamma i}(s,p,k,s_1, s_2)u_\alpha(p,s_1)u_\beta(k,s_2)\,e^{-ip\cdot y} +$$

$$+ d_{\gamma i}(s, p, -k, s_1, s_2)^{\dagger}v_\alpha(p,s_1)v_\beta(-k,s_2)\,e^{ip\cdot y}]\}$$

where $-k = (k^0, -\mathbf{k})$. This integration and similar integrations lead to the form of particle and gambol creation/annihilation operators seen below.

---

[11] See S. Blaha, Il Nuovo Cimento **49A**, 35 (1979) for one coordinate system, PseudoQuantum fermions.

3.2 Gambol Creation/Annihilation Operators

We define the gambol operators with

$$b^{1/s}{}_{g\gamma i}(s, q, s_2) = \Sigma_{s_1} (\int d^{r-1}pN'(p,q))^{½} b^{1/s}{}_{\gamma i}(s, p, q, s_1, s_2)$$

(3.17)

$$b^{1/s}{}_{g\gamma i}(s, q, s_2)^\dagger = \Sigma_{s_1} (\int d^{r-1}pN'(p,q))^{½} b^{1/s}{}_{\gamma i}(s, p, q, s_1, s_2)^\dagger$$

where s is the fractionation and $N'(p,q) = (2\pi)^{r-1}N(p,q)$. Then we define the gambol field operator anti-commutators with the anti-commutation relations:

$$\{b^{1/s}{}_{g\gamma i}(s, q, s_2), b^{1/s'}{}_{g\gamma' j}(s', q', s_2')^\dagger\} = \tag{3.18}$$
$$= \Sigma_{s_1} \Sigma_{s_1'} (\int d^{r-1}pN'(p,q))^{½}(\int d^{r-1}p'N'(p',q))^{½} \{b^{1/s}{}_{\gamma i}(s,p,q,s_1,s_2), b^{1/s'}{}_{\gamma' j}(s', p', q', s_1', s_2')^\dagger\}$$
$$= (\int d^{r-1}pN'(p,q))^{½}(\int d^{r-1}p'N'(p',q))^{½}(1 - \delta_{ij})\, \delta_{s,s'}\, \delta_{s_2 s_2'}\, \delta_{\gamma,\gamma'}\delta^{r-1}(\mathbf{p} - \mathbf{p}')\delta^{r-1}(\mathbf{q} - \mathbf{q}')\, U_g(\varepsilon(s,q))$$
$$= \int d^{r-1}p\, N'(p,q)\ (1 - \delta_{ij})\, \delta_{s,s'}\, \delta_{s_2,s_2'}\, \delta_{\gamma,\gamma'}\delta^{r-1}(\mathbf{p} - \mathbf{p}')\delta^{r-1}(\mathbf{q} - \mathbf{q}')\, U_g(\varepsilon(s,q))$$
$$= \int d^{r-1}p\, N'(p',q)\, \delta^{r-1}(\mathbf{p} - \mathbf{p}')\, (1 - \delta_{ij})\, \delta_{s,s'}\, \delta_{s_2,s_2'}\, \delta_{\gamma,\gamma'}\delta^{r-1}(\mathbf{q} - \mathbf{q}')\, U_g(\varepsilon(s,q))$$
$$= (1 - \delta_{ij})\, \delta_{s,s'}\delta_{s_2,s_2'}\, \delta_{\gamma,\gamma'}\delta^{r-1}(\mathbf{q} - \mathbf{q}')\, U_g(\varepsilon(s,q))$$

and

$$\{d^{1/s}{}_{g\gamma i}(s, q, s_2), d^{1/s'}{}_{g\gamma' j}(s', q', s_2')^\dagger\} = (1 - \delta_{ij})\, \delta_{s,s'}\delta_{s_2,s_2'}\, \delta_{\gamma,\gamma'}\delta^{r-1}(\mathbf{q} - \mathbf{q}')\, U_g(\varepsilon(s,q)) \tag{3.19}$$

where eqs. 2.22 specifies $U_g(\varepsilon(s,q))$ with m = M.

Eqs. 3.18 and 3.19 have fractional integrations that utilize

$$[(\int dp^{r-1}\, N'(p,q))^{½}]^2 = \int dp^{r-1}\, N'(p,q) \tag{3.20}$$

as in Riemann-Liouville integrals. Note γ and γ' are internal symmetry indices.

Thus particles and gambols have the same internal symmetries and spin. They differ in mass and momentum. The corresponding d type operators have similar anti-commutation relations. The other anti-commutators are zero:

$$\{b^{1/s}{}_{g\gamma i}(s, q, s_2)^\dagger, b^{1/s'}{}_{g\gamma' j}(s', q', s_2')^\dagger\} = 0 \tag{3.21}$$
$$\{d^{1/s}{}_{g\gamma i}(s, q, s_2)^\dagger, d^{1/s'}{}_{g\gamma' j}(s', q', s_2')^\dagger\} = 0$$
$$\{b^{1/s}{}_{g\gamma i}(s, q, s_2), b^{1/s'}{}_{g\gamma' j}(s', q', s_2')\} = 0$$
$$\{d^{1/s}{}_{g\gamma i}(s, q, s_2), d^{1/s'}{}_{g\gamma' j}(s', q', s_2')\} = 0$$

and the anti-commutators of b and d type operators are zero as well.

The gambol creation/annihilation operators above can be used to define gambol fermion quantum fields.

$$\psi_{g\gamma 1}(x) = \Sigma_p\, [b^{1/s}{}_{g\gamma 1}(s, p, s_1)\, f_p(x) + d^{1/s}{}_{g\gamma 1}(s, p, s_1)^\dagger f_p{}^*(x)] \tag{3.21a}$$
$$\psi_{g\gamma 2}(x) = \Sigma_p\, [b^{1/s}{}_{g\gamma 2}(s, p, s_1) f_p(x) + d^{1/s}{}_{g\gamma 2}(s, p, s_1)^\dagger\, f_p{}^*(x)]$$

3.3 Particle Creation/Annihilation Operators

These operators may be defined using the composite creation/annihilation of eq. 3.14 for particles of momentum p. We replace[12] the sum over s[13] with a sum over energies ε so as to take advantage of the normalization sum in eq. 2.21:[14]

$$b_{\gamma i}(p, s_2) = \int_0^\infty d\varepsilon\, \Sigma_{s_1} (\int d^{r-1}q\, N'(p,q))^{1/2}\, b^{1/s}{}_{\gamma i}(s, p, q, s_1, s_2) \qquad (3.22)$$

$$b_{\gamma i}(p, s_2)^\dagger = \int_0^\infty d\varepsilon\, \Sigma_{s_1} (\int d^{r-1}q\, N'(p,q))^{1/2}\, b^{1/s}{}_{\gamma i}(s, p, q, s_1, s_2)^\dagger$$

The anti-commutation relations are[15]

$$\begin{aligned}
&\{b_{\gamma i}(p, s_2), b_{\gamma' j}(p', s_2')^\dagger\} = \\
&= \int d\varepsilon \int d\varepsilon'\, \Sigma_{s_1} \Sigma_{s_1'} (\int d^{r-1}qN'(p,q))^{1/2} (\int d^{r-1}q'N'(p,q))^{1/2} \{b^{1/s}{}_{\gamma i}(s,p,q,s_1,s_2), b^{1/s'}{}_{\gamma' j}{}^\dagger(s',p',q',s_1',s_2')\} \\
&= \int d\varepsilon \int d\varepsilon' \Sigma_{s_1} \Sigma_{s_1'} (\int d^{r-1}qN'(p,q))^{1/2} (\int d^{r-1}q'N'(p,q))^{1/2} (1 - \delta_{ij})\, \delta_{s_1,s_1'}\, \delta_{s_2,s_2'} \delta_{ss'} \delta_{\gamma\gamma'} \delta^{r-1}(\mathbf{p} - \mathbf{p'}) \cdot \\
&\quad \cdot \delta^{r-1}(\mathbf{q} - \mathbf{q'}) U_g(\varepsilon(s,q)) \\
&= \int d^{r-1}q\, N'(p,q)\, (1 - \delta_{ij})\, \delta_{s_2,s_2'}\, \delta_{\gamma\gamma'} \delta^{r-1}(\mathbf{p} - \mathbf{p'}) \delta^{r-1}(\mathbf{q} - \mathbf{q'}) \int d\varepsilon\, U_g(\varepsilon(s,q)) \\
&= \int d^{r-1}q\, N'(p,q)\, \delta^{r-1}(\mathbf{q} - \mathbf{q'})\, (1 - \delta_{ij})\, \delta_{s_2,s_2'}\, \delta_{\gamma\gamma'} \delta^{r-1}(\mathbf{p} - \mathbf{p'}) \\
&= (1 - \delta_{ij})\, \delta_{s_2,s_2'}\, \delta_{\gamma\gamma'} \delta^{r-1}(\mathbf{p} - \mathbf{p'}) \qquad (3.23)
\end{aligned}$$

using eq. 3.21 with ε(s), and ε(s'). We make the s (and ε) sums into integrations for analytic convenience just as the discrete sum over hν in the black body Planck distribution derivation is similarly made continuous.

The other anti-commutation relations are:

$$\begin{aligned}
\{d_{\gamma i}(p, s), d_{\gamma' j}(p', s')^\dagger\} &= (1 - \delta_{ij})\, \delta_{s,s'}\, \delta_{\gamma\gamma'} \delta^{r-1}(\mathbf{p} - \mathbf{p'}) \qquad (3.24) \\
\{b_{\gamma i}(p, s), b_{\gamma' j}(p', s')\} &= 0 \\
\{d_{\gamma i}(p, s), d_{\gamma' j}(p', s')\} &= 0 \\
\{b_{\gamma i}(p, s)^\dagger, b_{\gamma' j}(p', s')^\dagger\} &= 0 \\
\{d_{\gamma i}(p, s)^\dagger, d_{\gamma' j}(p', s')^\dagger\} &= 0
\end{aligned}$$

and the anti-commutators of b and d type operators are zero.

The creation/annihilation operators above are those that appear in fermion quantum fields. Thus we may define a *particle* quantum field with

$$\begin{aligned}
\psi_{\gamma 1}(x) &= \Sigma_p\, [b_{\gamma 1}(p, s)\, f_p(x) + d_{\gamma 1}(p, s)^\dagger f_p{}^*(x)] \qquad (3.25) \\
\psi_{\gamma 2}(x) &= \Sigma_p\, [b_{\gamma 2}(p, s) f_p(x) + d_{\gamma 2}(p, s)^\dagger\, f_p{}^*(x)]
\end{aligned}$$

---

[12] We use s and ε interchangeably.

[13] The sum over s begins as a discrete sum over powers of 2 according to Limos. We replace it with a continuous value for s, which makes the transition to an integral over ε possible.

[14] Using a sum over ε supports eq. 3.13.

[15] The s, s' and $\delta_{ss'}$ terms are equivalent to ε energy terms due to the 1:1 relation of s and ε specified in eq. 2.23.

with internal symmetry index γ.

3.4 Map Between Gambol and Particle states

We now relate the particle wave function to sets of gambol wave functions by requiring a symmetry in the creation/annihilation operators:

$$b_{\gamma i}(s, p, q, s_1, s_2) = b_{\gamma i}(s, q, p, s_2, s_1) \qquad (3.26)$$
$$d_{\gamma i}(s, p, q, s_1, s_2) = d_{\gamma i}(s, q, p, s_2, s_1)$$

plus Hermitean conjugates. Then we may relate the gambol operators to the particle operators using eqs. 3.17 and 3.22:

$$b_{\gamma i}(p, s_2) = \int d\varepsilon \Sigma_{s_1} (\int d^{r-1}q N'(p,q))^{1/2}\, b^{1/s}{}_{\gamma i}(s, p, q, s_1, s_2) \qquad (3.27)$$
$$= \int d\varepsilon\, b^{1/s}{}_{g\gamma i}(s(\varepsilon), p, s_2)$$

Thus the particle operator is a sum of gambol operators. *The gambols have the same spin, momentum, and internal symmetry as the parent particle.* The gambol mass, $m_g = m/s$, appears in $\varepsilon(s,p)$. The gambol temperature is defined in terms of the particle mass: $kT_g = 0.0785m$. (See Blaha (2023e).)

These features may be seen in the form of $U_g(\varepsilon(s, p))$ in eqs. 2.5 and 2.23. Each term in eq. 3.27 defines an operator for a set of gambols from which one is projected in a particle interaction. The particle interacts via a sum of $b^{1/s}{}_{g\alpha i}(s, p, s_2)$ gambol interactions.

The particle's Green's function has the momentum space form:

$$G(p) = \int d\varepsilon\, U_g(\varepsilon(s, p))/\gamma\cdot p \qquad (3.28)$$

$$= 1//\gamma\cdot p$$

3.5 Particle-Gambol States

A one particle state is

$$|1\rangle = b_{\gamma 2}(p, s_2)^\dagger|0\rangle \qquad (3.29)$$

It is equivalent to a sum of sets of gambols for all values of s:

$$|g\, s, p, s_2\rangle = \Sigma_s\, b^{1/s}{}_{g\gamma 2}(s, p, s_2)^\dagger|0\rangle = |1\rangle \qquad (3.30)$$

by eq. 3.27.

Thus a particle state is equivalent to a superposition of gambol states. If we project a gambol representing a set of gambols, we find

$$\langle g\, s', p', s_2'|1\rangle = \delta_{s,s'}\delta_{s_2,s_2'}\delta^{r-1}(\mathbf{p} - \mathbf{p'})U_g(\varepsilon(s, p)) \qquad (3.31)$$

Thus a particle state is a probabilistic superposition of gambol states. We now have a quantum field theory for a particle – gambol configuration. Comparing this theory with the phenomenological cases studied in Blaha (2023e) we see a perfect match with that phenomenology.

3.6 Individual Gambols

The above formalism projects a gambol from a set of s gambols for fractionation s. The individual gambols in a set have identical quantum numbers. But the sets are defined as conceptual assemblages of s gambols due to a fractionation of a particle into sets of gambols. A gambol does not exist[16] until it is projected from an assemblage by a projection operator[17] $b^{1/s}{}_{g\alpha 2}(s, p, s_1)^\dagger$. Then it may interact.

3.7 Gambol Based S-Matrix

In Blaha (2023e) we defined the S-Matrix for gambol based hadron scattering. We calculate the interaction S-matrix for gambol-gambol scattering using gambol masses $m_g = m/s$ and four momenta $p_g = p/s$ but otherwise using the spins and internal symmetry features of each parent hadron. We treat the gambols similarly to their free hadron parents but with gambol masses and momenta.

> The S-matrix for a process is the product of gambol probability distributions and an S-matrix S_g for gambol-gambol scattering that has the same form as that of the parent particles (same internal symmetries and spins) but with gambol masses and momenta. S_g is specified to the lowest perturbative order. *Thus gambol probability distributions enable the S-matrix to be determined from a low order S-matrix S_g.*

We consider the example of two hadron scattering since it shows all the features of gambol based hadron scattering. We denote the low order gambol S-matrix for the scattering of two gambols into two gambols with

$$S_g(p_{g1}, m_{g1}, p_{g2}, m_{g2}, p_{g3}, m_{g3}, p_{g4}, m_{g4}) \tag{3.32}$$

for i = 1, 2, 3, 4. We use the interpolation expression of eq. 2.5 for the gambol energy.

The resulting hadron-hadron differential cross section is

$$d\sigma/d\Omega = N \int ds_1 ds_2 ds_3 ds_4 U_{g1}(\varepsilon_1(s_1,p_1))U_{g2}(\varepsilon_2(s_2,p_2))U_{g3}(\varepsilon_3(s_3,p_3))U_{g4}(\varepsilon_4(s_4,p_4))S_g^\dagger S_g \tag{3.33}$$

where

$$p_{gi} = p_i/s_i \tag{3.34}$$

$m_{gi} = m_i/s_i$ bare gambol mass and
$U_{gi}(\varepsilon_i(s_i, p_i))$ is the distribution for the ith particle for gambol energy ε_i. It differs from particle to particle.

---

[16] One may define an extension of the theory with creation/annihilation operators for individual gambols using operators such as $b^{1/S}{}_{g\gamma 2}(s, p, s_1)^\dagger$. Such an extension does not seem to have a purpose due to the nature of the formalism. A projection suffices for the purpose of quantum field theory calculations as below in section 3.7. It also avoids the issue of a conflict with the Pauli Exclusion Principle. Gambols in a set are not Real. Their projections are real.

[17] The subscript 2 is due to our PseudoQuantum Theory of fields.

and where N is a normalization.

We chose to specify the sets of four individual gambols of each of the four hadrons at the same fractionation s_i. Eq. 3.34 expresses the hadronic interaction as a gambol interaction where only the masses and momenta differ from those of their parent individual hadrons. Fig. 3.1 gives the example of $\pi^- K^0 \rightarrow p\ \Lambda^0$.

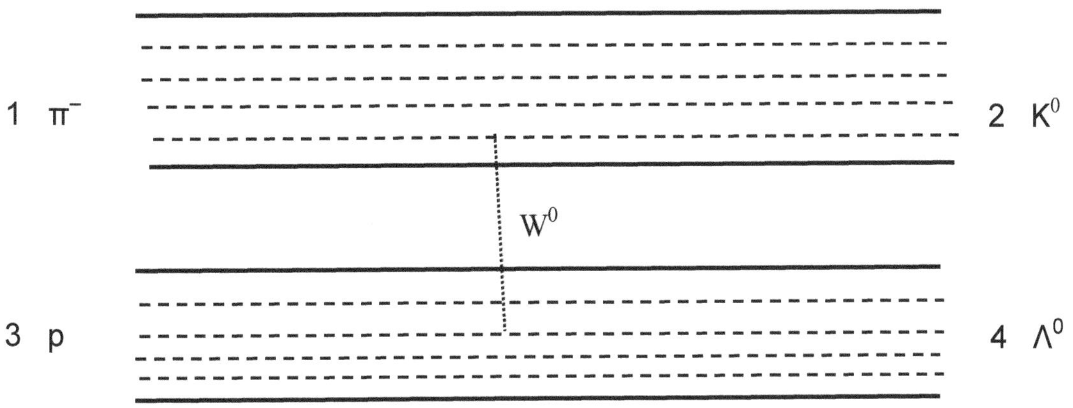

Figure 3.1. The reaction $\pi^- + p \rightarrow \Lambda^0 + K^0$ due to the exchange of a W^0 is symbolized for gambols. The integers denote the factors in eq. 3.33.

4. Universe – Gambol Quantum Fields

In chapter 6 we will develop a gambol model for universe Hubble expansion. In chapter 7 we will show that we could derive universe confinement in a manner like quark confinement within a containing Megaverse. In chapter 8 we develop a gambol based model for S_8 clumping in an evolving universe. In chapter 11 we develop a model for the distribution of universes of the ten HyperCosmos spaces based on the Planckian distribution.

These studies support the treatment of a universe as a type of particle within a Megaverse. (We view our universe as having a Cayley-Dickson n = 3 space. We treat the Megaverse as having an n = 4 space in Cosmos Theory. See Fig. 10.1 for the HyperCosmos set of spaces.

We now consider a quantum field theory where a universe is a particle with a quantum field representation in the Megaverse. The universe is assumed to have a gambol representation similar to the gambol representation of particles described in chapter 3. We will take advantage of the similarity to construct the universe – gambol quantum field theory. Following the particle-gambol discussion of chapter 3 we define[18] two coordinates PseudoFermions with the form

$$\psi_{i\alpha\beta u}(y, z) \tag{4.1}$$

where y and z are independent coordinates in r space-time dimensions, where i = 1, 2 labels the PseudoQuantum[19] fields, α and β are spinor indices, and the subscript "u" indicates a universe quantum field within an N = 6 Cosmos Megaverse.

4.1 Universe and Gambol Quantum Fields

We begin[20] by defining a free universe PseudoFermion PseudoQuantum Lagrangian with two wave functions ψ_{1u} and ψ_{2u} that are functions of two sets of coordinates in r space-time dimensions, y and z,

$$\begin{aligned}\mathscr{L} = \overline{\psi}_{2\alpha\beta u}[-M^{-1}\gamma_{y\alpha\kappa}{}^{\mu}\cdot\partial/\partial y^{\mu}\ \gamma_{z\beta\lambda}{}^{\nu}\partial/\partial z^{\nu} - M]\psi_{1\kappa\lambda u} + \\ + \overline{\psi}_{1\alpha\beta u}[-M^{-1}\gamma_{y\alpha\kappa}{}^{\mu}\cdot\partial/\partial y^{\mu}\ \gamma_{z\beta\lambda}{}^{\nu}\partial/\partial z^{\nu} - M]\psi_{2\kappa\lambda u}\end{aligned} \tag{4.2}$$

---

[18] Again we follow the conventions of Bjorken (1965) with the $g^{\mu\nu}$ metric (1, -1, -1, -1, -1, -1).

[19] Appendix C of Blaha (2016f) presents a first quantized PseudoQuantum theory CQ Mechanics that embodies both classical and quantum theory. **This theory is the non-relativistic quantum mechanics limit of the PseudoFermion theory developed here.** CQ Mechanics has two sets of coordinates that combine to create a generalization of conventional Quantum Mechanics. It has applications in a generalized Feynman path integral formalism, a generalized Schrödinger equation, a generalized Boltzmann equation, the Fokker-Planck equation, a generalized approach to quantum and classical chaos, and to quantum entanglement as well as semi-quantum entanglement. Our "Pseudo" formalisms apply to both Quantum Field Theory and Quantum Mechanics. In these applications there is a clear almost continuous transition between the quantum to the classical sectors.

[20] This development is similar to that of chapter 3.

where γ_y^μ and γ_z^μ are Dirac matrices for y and z coordinates respectively and y and z are coordinates in the r = 6 dimension Megaverse space-time, M is the mass, and

$$\overline{\psi}_{i\alpha\beta u} = \psi_{i\kappa\lambda u}{}^{\dagger}\, \gamma_y{}^{0}{}_{\kappa\alpha}\, \gamma_z{}^{0}{}_{\lambda\beta} \tag{4.2a}$$

for i = 1, 2 where the Dirac matrix subscripts y and z indicate they are associated with the y and z coordinates. The indices α and β are spinor indices.

The equations of motion are

$$[-M^{-1}\gamma_y{}^{\mu}\cdot\partial/\partial y^{\mu}\, \gamma_z{}^{\nu}\partial/\partial z^{\nu} - M]\psi_{1u} = 0 \tag{4.2b}$$
$$[-M^{-1}\gamma_y{}^{\mu}\cdot\partial/\partial y^{\mu}\, \gamma_z{}^{\nu}\partial/\partial z^{\nu} - M]\psi_{2u} = 0 \tag{4.2c}$$

We define subsidiary equations of motion

$$[i\gamma_y{}^{\mu}\cdot\partial/\partial y^{\mu} - M]\psi_{ju} = 0 \tag{4.2d}$$
$$[i\gamma_z{}^{\nu}\partial/\partial z^{\nu} - M]\psi_{ju} = 0 \tag{4.2e}$$

for j = 1, 2. Eqs. 4.2d and 4.2e imply eqs. 4.2b and 4.2c.

One conjugate momentum is (with two spinor indices α, β)

$$\pi_{y1\alpha\beta u} = \partial\mathscr{L}/\partial(\partial\psi_{1\alpha\beta u}/\partial y^0) = -M^{-1}(\gamma_z{}^{\nu}\partial/\partial z^{\nu}\gamma_z{}^{0}\psi_{2u}{}^{\dagger})_{\alpha\beta} = (\psi_{2u}{}^{\dagger}\gamma_z{}^{0})_{\alpha\beta} \tag{4.3}$$

after partial integrations (with surface terms having the value zero) of[21]

$$L = \int d^r y \int d^r z\, \mathscr{L}$$

using the subsidiary equation of motion:

$$\gamma_z{}^{\nu}\partial/\partial z^{\nu}\, \psi_{2u}{}^{\dagger} = -M\psi_{2u} \tag{4.4}$$

Similarly the other conjugate momenta are

$$\pi_{z1\alpha\beta u} = \partial\mathscr{L}/\partial(\partial\psi_{1\alpha\beta u}/\partial z^0) = -M^{-1}(\gamma_y{}^{\nu}\partial/\partial y^{\nu}\gamma_y{}^{0}\psi_{2u}{}^{\dagger})_{\alpha\beta} = (\psi_{2u}{}^{\dagger}\gamma_y{}^{0})_{\alpha\beta} \tag{4.5}$$
$$\pi_{y2\alpha\beta u} = \partial\mathscr{L}/\partial(\partial\psi_{2\alpha\beta u}/\partial y^0) = -M^{-1}(\gamma_z{}^{\nu}\partial/\partial z^{\nu}\psi_{1u}{}^{\dagger})_{\alpha\beta} = (\psi_{1u}{}^{\dagger}\gamma_z{}^{0})_{\alpha\beta} \tag{4.6}$$
$$\pi_{z2\alpha\beta u} = \partial\mathscr{L}/\partial(\partial\psi_{2\alpha\beta u}/\partial z^0) = -M^{-1}(\gamma_z{}^{\nu}\partial/\partial y^{\nu}\psi_{1u}{}^{\dagger})_{\alpha\beta} = (\psi_{1u}{}^{\dagger}\gamma_y{}^{0})_{\alpha\beta} \tag{4.7}$$

using the subsidiary equations of motion:

$$\gamma_z{}^{\nu}\partial/\partial z^{\nu}\, \psi_{1u}{}^{\dagger} = -M\psi_{1u}{}^{\dagger} \tag{4.8}$$
$$\gamma_y{}^{\nu}\partial/\partial y^{\nu}\, \psi_{1u}{}^{\dagger} = -M\psi_{1u}{}^{\dagger} \tag{4.9}$$

---

[21] Note y and z are in the same space-time (coordinate system.)

$$\gamma_y{}^{\nu}\partial/\partial y^{\nu}\,\psi_{1u}{}^{\dagger} = -M\psi_{1u}{}^{\dagger} \qquad (4.10)$$

We define new momenta to preserve y – z symmetry.

$$\pi_{1\alpha\beta u} = (\pi_{y1u}\gamma_z{}^0)_{\alpha\beta} = (\pi_{z1u}\gamma_y{}^0)_{\alpha\beta} = \psi_2{}^{\dagger}{}_{\alpha\beta u} \qquad (4.11)$$
$$\pi_{2\alpha\beta u} = (\pi_{y2u}\gamma_z{}^0)_{\alpha\beta} = (\pi_{z2u}\gamma_y{}^0)_{\alpha\beta} = \psi_1{}^{\dagger}{}_{\alpha\beta u} \qquad (4.12)$$

The form of the conjugate momenta implies the only non-zero equal time anti-commutators are:[22]

$$\{\pi_{j\alpha\beta u}(\mathbf{y}, y^0, \mathbf{z}, z^0),\ \psi_{i\kappa\lambda u}(\mathbf{y'}, y^0, \mathbf{z'}, z^0)\} = \{\psi_{j\alpha\beta u}{}^{\dagger}(\mathbf{y}, y^0, \mathbf{z}, z^0),\ \psi_{i\kappa\lambda u}(\mathbf{y'}, y^0, \mathbf{z'}, z^0)\}$$

$$= (1 - \delta_{ij})\,\delta_{\alpha\kappa}\,{}_{\beta}\delta_{\beta\lambda}\,\delta^{r-1}(\mathbf{y} - \mathbf{y'})\delta^{r'-1}(z - \mathbf{z'}) \qquad (4.13)$$

for i, j = 1, 2.

If we introduce fractionation s where the particle is fractionated to s gambols, and internal symmetry index γ, the free PseudoFermion wave function has the form:

$$\psi_{i\alpha\beta\gamma u}(y,z) = \sum_{s}\sum_{s_1}\sum_{s_2} \int dp^{r-1}\int dq^{r-1}\, N(p, q)[b_{\gamma iu}(s, p, q, s_1, s_2)u_\alpha(p,s_1)u_\beta(q,s_2)\exp(-ip\cdot y - iq\cdot z) +$$

$$+ d_{\gamma iu}(s, p, q, s_1, s_2)^{\dagger} v_\alpha(p,s_1)v_\beta(q,s_2)\exp(ip\cdot y + iq\cdot z)] \qquad (4.14)$$

plus Hermitean conjugates for i = 1, 2 where N(p, q) is a normalization factor.

The composite creation and annihilation operators satisfy the anti-commutation relations:

$$\{b_{\gamma iu}(s, p, q, s_1, s_2), b_{\gamma' ju}(s', p', q', s_1', s_2')^{\dagger}\} = \{d_{\gamma iu}(s, p, q, s_1, s_2), d_{\gamma' ju}(s', p', q', s_1', s_2')^{\dagger}\} =$$
$$= (1 - \delta_{ij})\,\delta_{\gamma\gamma'}\,\delta_{s,s'}\delta_{s_1,s_1'}\,\delta_{s_2,s_2'}\,\delta^{r-1}(\mathbf{p} - \mathbf{p'})\delta^{r'-1}(\mathbf{q} - \mathbf{q'})\,U_g(\varepsilon(s,q)) \qquad (4.15)$$

where U_g is given by eqs. 2. 22.

The other composite anti-commutation operators are zeroes.

$$\{b_{\gamma iu}(s, p, q, s_1, s_2), b_{\gamma' ju}(s', p', q', s_1', s_2')\} = 0 \qquad (4.16)$$
$$\{b_{\gamma iu}(s, p, q, s_1, s_2)^{\dagger}, b_{\gamma' ju}(s', p', q', s_1', s_2')^{\dagger}\} = 0$$
$$\{d_{\gamma iu}(s, p, q, s_1, s_2), d_{\gamma' ju}(s', p', q', s_1', s_2')\} = 0$$
$$\{d_{\gamma iu}(s, p, q, s_1, s_2)^{\dagger}, d_{\gamma' ju}(s', p', q', s_1', s_2')^{\dagger}\} = 0$$

and anti-commutators of b and d type operators are zero as well.

4.2 Universe and Gambol Operators

The creation/annihilation operators may be divided into universe and gambol operators as in sections 3.2 – 3.5 where the particle operators are now universe creation and annihilation operators while the form of gambol operators is essentially the same.

---

[22] See S. Blaha, Il Nuovo Cimento **49A**, 35 (1979) for one coordinate system, PseudoQuantum fermions.

4.3 Universe Planckian Distribution

In Blaha (2023e), and in chapter 2, we developed a formulation of the gambol Planckian distribution. It was based on the statistical Einstein derivation of the Planck distribution together with a change in the energy ε from a massless photon basis to a particle mass basis for use in particle theories.

We will now extend that development to the case of gambols within universes as developed above. The first question is the framework of the derivation of the Planckian distribution. We address this issue by expanding on the Einstein derivation.

4.3.1 Einstein-like Derivation of the Planckian Distribution

Einstein derived the Planck distribution following a statistical approach. Einstein's[23] derivation starts with the concept of quantum jumps between energy states $E_1 < E_2$ due to a radiation density u.

In the case of universes we may define the states of a universe in several ways:

1. In states (stages) of different size if one wishes to develop a distribution for universe Hubble expansion. Initially the sizes can be viewed as discrete. Then they can be taken to be continuous. This case is addressed in chapter 6.

2. In states based on the changing, as time progresses, average mass-energy content of "clusters" of mass-energy such as galaxy clusters, galaxies, black holes, and so on. As time progresses the average mass-energy of clusters will change due to the evolution of the universe from a point (the Big Bang). This case will be addressed here. ***The purpose is to possibly explain the large galaxies and black holes that have been found experimentally at the beginning of time as a statistical effect.***

3. In states based on the clumping of mass-energy into regions of large and small density. This case will be addressed in chapter 8 in the discussion of the significance of the S_8 clumping index.

We now consider case 2 where the average size of mass-energy clusters changes with time. We assume the possible states for the average are initially discrete and undergo "quantum" jumps. Then we progress to the continuous case.

There are three ways a quantum jump occurs: spontaneous jumps

- with probability S_{12} for a transition from E_2 to E_1 with a quantum emission;
- with probability $A_{21}u$ absorption of a quantum giving a jump from E_1 to E_2 where u is the energy density;

---

[23] A. Einstein, Verhandl, deut. Physic. Ges. **18**, 318 (1916); Physik. Z. **18**, 121 (1917); J. W. M. Dumond and E. R. Cohen, Revs. Mod. Physics **25**, 691 (1953).

- with probability $I_{12}u$ for induced emission of a quantum giving a jump from E_1 to E_2 where u is the energy density

The A_{21} and I_{12} transitions appear with a "u" energy density factor since they are dependent on the available energy density. Absorption and emission are directly based on the availability of mass-energy.

The probabilities that states 1 and 2 are in thermal equilibrium proportional to $\exp(-E_1/kT)$ and $\exp(-E_2/kT)$ respectively. Thus

$$(S_{12} + I_{12}u) \exp(-E_2/kT) = A_{21}u \exp(-E_1/kT) \tag{4.17}$$

expresses the balance of transitions. The constants S_{12}, I_{12} and A_{21} are properties of the states and independent of the density u. At high temperature $S_{12} \ll I_{12}u$. This implies

$$I_{12} = A_{21} \tag{4.18}$$

The transitions: $1 \rightarrow 2$ and $2 \rightarrow 1$ induced by the electromagnetic field are the same.

The density u may then be calculated to be

$$u = (S_{12}/I_{12})/(\exp((E_2 - E_1)/kT) - 1) \tag{4.19}$$

Einstein's derivation assumes that the container and the radiation may coexist in a stable state and that transitions between states can be determined statistically. These conditions are true for a universe.

Eq. 4.19 is the basis of the particle – gambol distribution[24] U_g of eq. 2.22. We will use it to find the distribution in time of the average mass-energy value from the Big Bang to the present. The Planckian distribution may be viewed as giving the number of $\varepsilon(s)$ "quanta" of fractionation s.

4.4 Universe and Gambol Operators and States

The results of sections 3.2 – 3.7 apply to universes and universe gambols with no changes.

4.5 Time Evolution of Universe Mass-Energy Entities

There is interest in the clumping of mass-energy in the universe. The S_8 Clumping Index (chapter 8) is an example. Here we consider the related issue of entities within the universe, that we call clusters that have larger mass-energy densities than their surroundings. The clusters range over stars, star clusters, galaxies, galaxy clusters and superclusters.

We assume each cluster is a gambol with a mass-energy ε. Each cluster is characterized by a fractionation 1/s of the universe's total mass-energy. The universe

[24] Chapter 6 derives our Anti-Planckian distribution giving the Hubble parameter as a fuction of time. This distribution derives is derived under different assumptions. It is based on an alternate gambol basis for the universe. In the case of particles, having "instantaneous" interactions and a static inner nature, we find one gambol description is sufficient. In the case of universes which have a long time evolution there are several possible gambol models.

mass-energy is the sum of the mass-energies of the set of all clusters for all values of s. It is time dependent.

The average mass-energy of all clusters of fractionation 1/s counts the number[25] *of such clusters of each energy ε.* It is specified by our Planckian distribution E_ε:

$$E_\varepsilon = U_g(\varepsilon(s)) = 15\, N\, (\pi kT)^{-4}\, \varepsilon^3/(e^{\varepsilon/kT} - 1) \tag{4.20}$$

where N has the dimensions GeV^2/c^4 to make E_ε have the correct dimension GeV/c^2.

We previously defined the individual gambol energy parameter with

$$\varepsilon(s) = [(M_g\, s/M_u + 1)/(s + 1)]E \tag{4.21}$$

where s is the fractionation parameter, M_g is the gambol mass, M_u is the total mass-energy of the universe (including Dark energy and matter), and E is the kinetic energy of the universe. The only reasonable current choice is $E = M_u$ indicating the universe is not in motion.

The distribution of gambols varies in time due to gravitation and interactions which are implicitly confined by the boundaries of the universe. We now introduce time by defining ε in terms of time:

$$s = 1/(bt) \tag{4.22}$$

resulting in

$$\varepsilon(t) = (M_g + btM_u)/(bt + 1) \tag{4.23}$$

Now E_g, the average energy for the ε mass-energy gambols, varies with time. One expects that the average total mass-energy of gambols will be very small at times after the Big Bang when the fractionation s is very large. Then as time progresses the average energy of entities increases and s decreases.

The Gambol Model[26] of Universe Mass-Energy Entities is based on a universe split by decreasing fractionation values as the universe evolves. At each time, viewing time as discrete for the purpose of discussion, ε(t) and E_g change just as the black body Planck distribution for photons changes with photon energy hν. The mass-energy size of each gambol is initially a 1/s part of the universe mass-energy where s is a power of 2. We then make s continuous.

In chapter 8 we will relate E_ε to the S_8 Clumping Index

$$S_8 = N_{S8}\, E_\varepsilon/M_u$$

---

[25] The number of clusters of fractionation s will be shown later to vary in time according to E_g. The model is analogous to the Planck black body distribution. For each photon of energy hν there are a certain number of photons. This number is specified by the black body distribution value for hν. Similarly, the average number of gambols, which we map to an energy, is specified by E_g.

[26] This somewhat sensitive calculation is performed in double precision using Excel. The calculations in Blaha (2023e) are also performed in double precision using Excel. Double precision mathematics was also required in the author's calculation of the Fine Structure Constant since it is known to thirteen decimal places. The calculation found the exact known result.

where N_{S8} is a normalization,

4.6 Inputs for the Gambol Model of Mass-Energy Entities

We now specify necessary inputs for the Model. We bring in time dependence by defining a constant[27] b with the empirical value[28]

$$b = \; = 8.67 \times 10^{-19} \text{ sec}^{-1} \tag{4.24}$$

for use in transforming s into an equivalent time by eq. 4.22.

We define the universe *gambol* temperature T_{gu} of the universe with

$$kT_{gu} = 0.0785\, M_u(s + 1) \; = 0.0785\, M_u\, [1/(bt) + 1] \tag{4.25}$$

Eq. 4.25 uses almost the same definition as for particles in Blaha (2023e). The universe gambol temperature is slightly modified[29] from its equation for the gambol temperature for quarks, leptons, and hadrons in Blaha (2023e):

kT_{gu} may be expressed as a function of ε using

$$bt \; = [M_g - \varepsilon]/(\varepsilon - M_u) \tag{4.26}$$

$$kT_{gu} = \; = 0.0785\, M_u\, [(M_u - \varepsilon)/(\, \varepsilon - M_g) + 1] \tag{4.27}$$

Thus we can view $U_g(\varepsilon(s)) \equiv U_g(\varepsilon(t))$ as strictly a function of ε.

4.6.1 Determination of Constants[30]

The **total** mass-energy[31] of the universe has been *estimated* to be

$$M = 1.712 \times 10^{81} \text{ GeV/c}^2 = E \tag{4.28}$$

Our mass for gambols is[32]

$$M_g = M_u/s = btM_u \tag{4.29}$$

The age of the universe has been estimated to be

$$t_{NOW} = \; 13.787 \text{ billion years} = 4.35 \times 10^{17} \text{ sec} \tag{4.30}$$

The universe temperature now is

---

[27] Chapter 12 suggests that b might indicate the "end" time of the universes when it ceases to expand further.

[28] This value was used in Blaha (2023e).

[29] The insertion of 1 in this equation enables the gambol temperature to be a non-zero constant as s → 0. The s = 0 value is outside the effective fractionation range; so we physically want $s_{effective} = 1$ as the least allowed value of s for fractionation. We extend s to zero in calculations and make it continuous to have analytic computations. The same considerations apply in eq. 4.22a where s + 1 appears.

[30] The extremely large size of some variables and the extremely small size of other variables causes the calculation to be sensitive to the parameters.

[31] This value includes normal mass and energy plus Dark Energy and Dark Matter.

[32] We change this expression to $M_g = M_u/1024$ later.

$$kT_{NOW} = k\,2.72548\text{ K} = 2.126 \times 10^{8}\text{ GeV/c}^2 \qquad (4.31)$$

The present physical T is related to the present T_{gu} by

$$kT_{physNOW} = 5.97 \times 10^{-73}\,kT_{guNOW} \qquad (4.32)$$

The normalization N is chosen to make the $t_0 = b^{-1}$ average mass-energy $U_g(\varepsilon(t_0))$ equal to M_u:[33]

$$N = 2.15 \times 10^{162}\text{ GeV}^2/c^4 \approx M_u^{\,2} \qquad (4.33)$$

Note: $N = M_u^{\,2} = 2.9 \times 10^{162}\text{ GeV}^2/c^4$ is a very good approximation to N. It has the correct dimensions agreeing with E_ε's dimension GeV/c^2.

4.7 The Universe Gambol Mass-Energy Model Results

The Planckian distribution is based on the Planck distribution. The Planck distribution gives the energy of a state as a function of the energy ε – the energy of a quantum. In analogy, we defined a model that gives the energy of a fractionation in terms of the "gambol quantum" ε. We use the one-to-one relation between ε and time t in U_g, to determine the time distribution of the average Mass-Energy of the universe's sets of gambols in time.. As time progresses we find the average Mass-Energy grows.

At t = 0 (the Big Bang point) we expect the average Mass-Energy to be zero. There is an infinity of highly fractionated gambols.

The calculation involves both extremely large and extremely small quantities. It is constrained by the framework of gambol particle equations and parameter values found in *particle* (quark, lepton and hadron) calculations

Figs. 4.1 and 4.2 show the trend of average mass-energy with time. Notice the rapid growth in average mass-energy in early times (Fig. 4.1) suggestive of the rapid growth of galaxies and clusters found experimentally. For very large times, in the future, the average mass-energy approaches the mass-energy of the universe: $M = 1.712 \times 10^{81}\text{ GeV/c}^2$. Then $s \approx 1$.

Fig. 4.2 shows a more moderate growth in recent times. Thus the average mass-energy, which computes an average for the sum of states of energy ε(s), grows fairly rapidly in times near the Big Bang and relatively more moderately in recent times. It relates to S_8, which describes the "blanketing" of the universe by the combined mass-energy of all universe entities.

4.8 Some Clustering of Mass-Energy

There is experimental evidence for very large mass-energy clustering:

- The Virgo Supercluster comprising 100,000 galaxies
- The Great Attractor "uniting" 400 galaxies

---

[33] In this case s = 1 and the universe factorization thus gives a gambol equivalent to the entire universe.

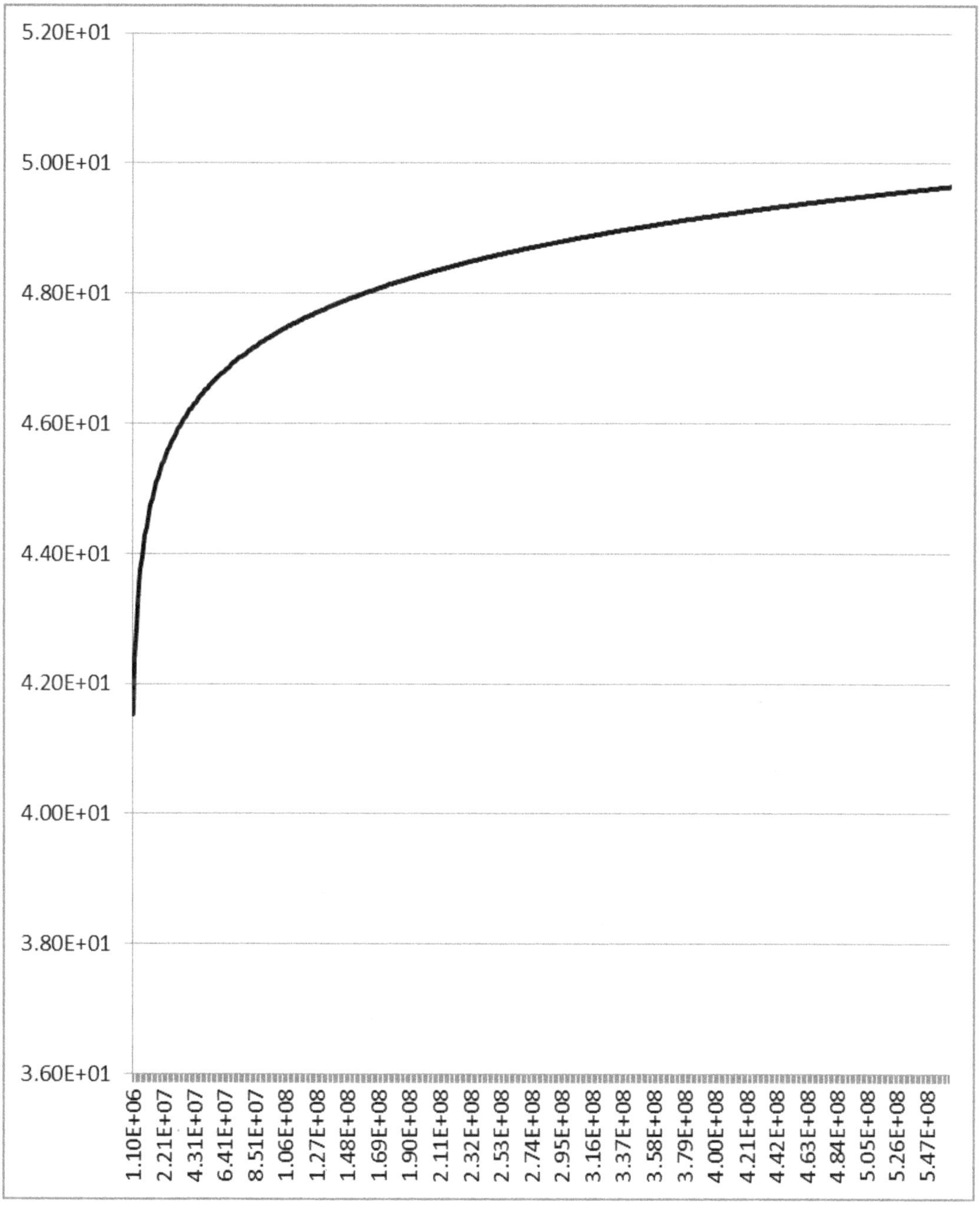

Figure 4.1. Early time, logarithm base 10 of the average mass-energy as a function of time measured in seconds from the Big Bang.

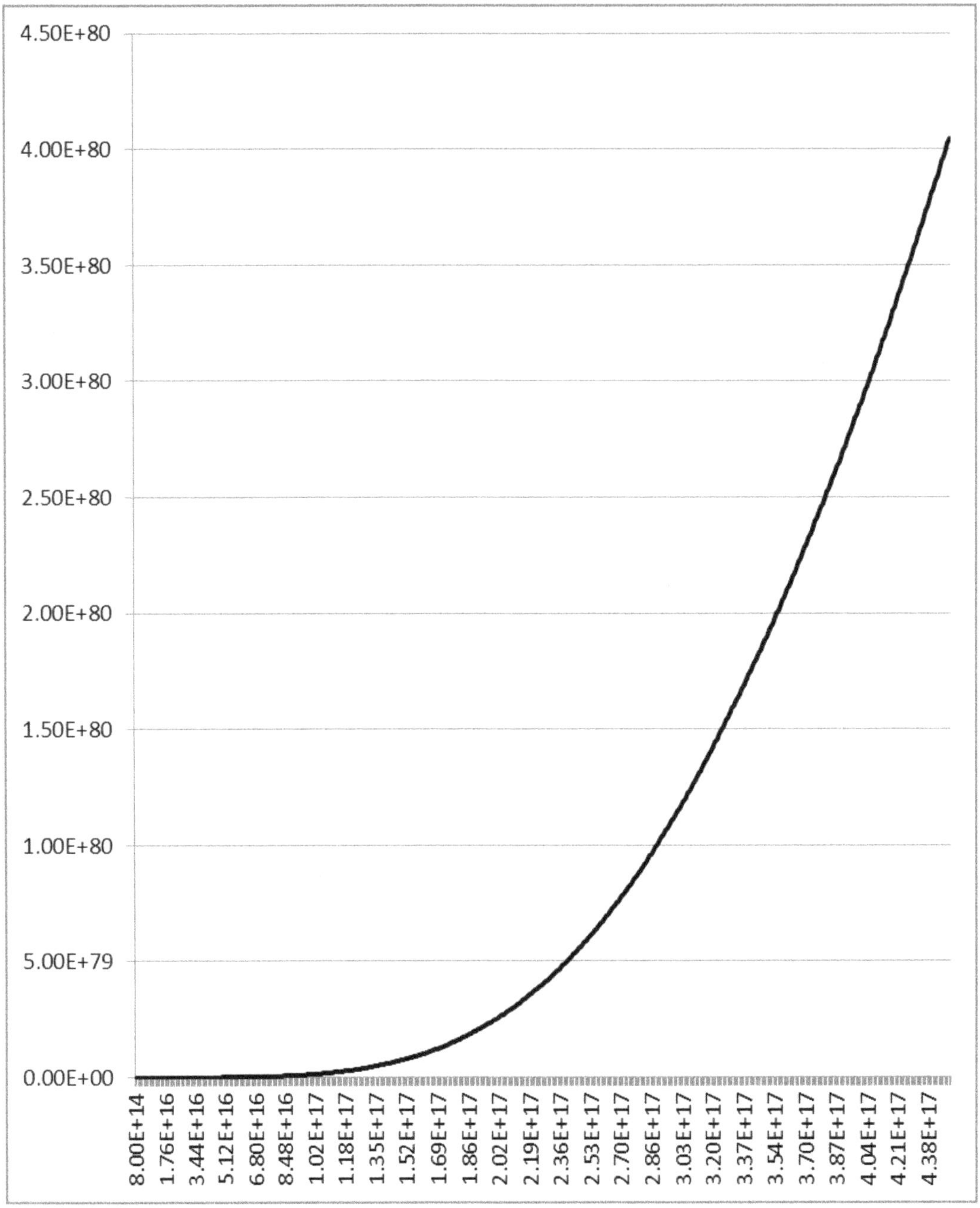

Figure 4.2 Recent time, average mass-energy in GeV/c^2 as a function of time measured in seconds from the Big Bang.

5. Universes as Gambols in the Megaverse

We think of universes such as our universe as "substantial." However this point of view may be an illusion. It is possible to think of universes as gambols of varying probabilities within a Megaverse. Then they become probabilistic threads of Reality – Real in one sense – yet probabilistic in nature.[34]

Studies[35] in this, and our earlier, books support the treatment of our universe[36] as a type of particle within a six space-time dimension Megaverse. (We view the universe as having a Cayley-Dickson n = 3 space with four space-time dimensions. We treat the Megaverse as having an n = 4 space in Cosmos Theory.)

We now consider a quantum field theory[37] for a Megaverse particle[38] composed of universe gambols. The universe is assumed to have an internal gambol representation similar to the gambol representations described in chapters 3 and 4. We will take advantage of the similarity to construct a Megaverse particle – universe gambol representation where the gambols are universes.

Following the particle-gambol discussion of chapter 3 we define two coordinates PseudoFermions for a Megaverse quantum field containing universe gambols:

$$\psi_{i\alpha\beta M}(y, z) \tag{5.1}$$

where y and z are independent coordinates in r' = 6 and r = 4 space-time dimensions respectively, where i = 1, 2 labels the PseudoQuantum fields, α and β are spinor indices, and the subscript "M" is the mass of a Megaverse quantum field containing universe gambols. The y coordinates are for the Megaverse particle; the z coordinates are for the universe gambols.

5.1 Universe Gambol Quantum Fields

We begin by defining a free Megaverse PseudoFermion PseudoQuantum Lagrangian with two wave functions ψ_{1u} and ψ_{2u} that are functions of two sets of coordinates in r' and r space-time dimensions, y and z,

---

34 We must remember Dr. Samuel Johnson who refuted theories of the insubstantial with a kick.

35 In chapter 6 we will develop a gambol model for universe Hubble expansion. In chapter 7 we will show that we could derive universe confinement like quark confinement within the containing Megaverse. In chapter 8 we develop a gambol based model for S8 clumping in an evolving universe.

36 We think of a universe as a mass-energy clump whose structure is governed by the dimension array of one of the ten spaces of the HyperCosmos.

37 We follow a similar derivation to the derivation in chapter 4.

38 The Megaverse quantum field resides in a subspace of its parent universe. See chapter 10 for details. Chapter 10 extends this line of thinking to posit the Cosmos is a chain of gambols from an n = 10 universe down to n = 3 universes. Chapter 11 develops a Planckian distribution for the set of universes of the ten HyperCosmos spaces. It suggests that there are 8 N = 7 HyperCosmos universes, one of which is our universe.

$$\mathscr{L} = \overline{\psi}_{2\alpha\beta M}[-M^{-1}\gamma_{y\alpha\kappa}{}^{\mu}\cdot\partial/\partial y^{\mu}\ \gamma_{z\beta\lambda}{}^{\nu}\partial/\partial z^{\nu} - M]\psi_{1\kappa\lambda M} + $$
$$+ \overline{\psi}_{1\alpha\beta M}[-M^{-1}\gamma_{y\alpha\kappa}{}^{\mu}\cdot\partial/\partial y^{\mu}\ \gamma_{z\beta\lambda}{}^{\nu}\partial/\partial z^{\nu} - M]\psi_{2\kappa\lambda M} \tag{5.2}$$

where $\gamma_y{}^{\mu}$ and $\gamma_z{}^{\mu}$ are Dirac matrices for y and z coordinates respectively, and y and z are coordinates in the r' = 6 and r = 4 dimension space-times respectively, M is the Megaverse quantum field mass, and

$$\overline{\psi}_{i\alpha\beta M} = \psi_{i\kappa\lambda M}{}^{\dagger}\ \gamma_y{}^{0}{}_{\kappa\alpha}\ \gamma_z{}^{0}{}_{\lambda\beta} \tag{5.2a}$$

for i = 1, 2 where the Dirac matrix subscripts y and z indicate they are associated with the y and z coordinates.

The equations of motion are

$$[-M^{-1}\gamma_y{}^{\mu}\cdot\partial/\partial y^{\mu}\ \gamma_z{}^{\nu}\partial/\partial z^{\nu} - M]\psi_{1M} = 0 \tag{5.2b}$$
$$[-M^{-1}\gamma_y{}^{\mu}\cdot\partial/\partial y^{\mu}\ \gamma_z{}^{\nu}\partial/\partial z^{\nu} - M]\psi_{2M} = 0 \tag{5.2c}$$

We define subsidiary equations of motion

$$[i\gamma_y{}^{\mu}\cdot\partial/\partial y^{\mu} - M]\psi_{jM} = 0 \tag{5.2d}$$
$$[i\gamma_z{}^{\nu}\partial/\partial z^{\nu} - M]\psi_{jM} = 0 \tag{5.2e}$$

for j = 1, 2. Eqs. 5.2d and 5.2e imply eqs. 5.2b and 5.2c.

One conjugate momentum is (with two spinor indices α, β)

$$\pi_{y1\alpha\beta M} = \partial\mathscr{L}/\partial(\partial\psi_{1\alpha\beta M}/\partial y^{0}) = -M^{-1}(\gamma_z{}^{\nu}\partial/\partial z^{\nu}\,\gamma_z{}^{0}\psi_{2M}{}^{\dagger})_{\alpha\beta} = (\psi_{2M}{}^{\dagger}\gamma_z{}^{0})_{\alpha\beta} \tag{5.3}$$

after partial integrations (with surface terms having the value zero) of

$$L = \int d^{r'}y \int d^{r}z\ \mathscr{L}$$

using the subsidiary equation of motion:

$$\gamma_z{}^{\nu}\partial/\partial z^{\nu}\ \psi_{2M}{}^{\dagger} = -M\psi_{2M} \tag{5.4}$$

Similarly the other conjugate momenta are

$$\pi_{z1\alpha\beta M} = \partial\mathscr{L}/\partial(\partial\psi_{1\alpha\beta M}/\partial z^{0}) = -M^{-1}(\gamma_y{}^{\nu}\partial/\partial y^{\nu}\,\gamma_y{}^{0}\psi_{2M}{}^{\dagger})_{\alpha\beta} = (\psi_{2M}{}^{\dagger}\gamma_y{}^{0})_{\alpha\beta} \tag{5.5}$$
$$\pi_{y2\alpha\beta M} = \partial\mathscr{L}/\partial(\partial\psi_{2\alpha\beta M}/\partial y^{0}) = -M^{-1}(\gamma_z{}^{\nu}\partial/\partial z^{\nu}\,\psi_{1M}{}^{\dagger})_{\alpha\beta} = (\psi_{1M}{}^{\dagger}\gamma_z{}^{0})_{\alpha\beta} \tag{5.6}$$
$$\pi_{z2\alpha\beta M} = \partial\mathscr{L}/\partial(\partial\psi_{2\alpha\beta M}/\partial z^{0}) = -M^{-1}(\gamma_z{}^{\nu}\partial/\partial y^{\nu}\,\psi_{1M}{}^{\dagger})_{\alpha\beta} = (\psi_{1M}{}^{\dagger}\gamma_y{}^{0})_{\alpha\beta} \tag{5.7}$$

using the subsidiary equations of motion:

$$\gamma_z^{\,v}\partial/\partial z^v\,\psi_{1M}^{\,\dagger} = -\,M\psi_{1M}^{\,\dagger} \tag{5.8}$$
$$\gamma_y^{\,v}\partial/\partial y^v\,\psi_{1M}^{\,\dagger} = -\,M\psi_{1M}^{\,\dagger} \tag{5.9}$$
$$\gamma_y^{\,v}\partial/\partial y^v\,\psi_{1M}^{\,\dagger} = -\,M\psi_{1M}^{\,\dagger} \tag{5.10}$$

We define new momenta to preserve y – z symmetry.

$$\pi_{1\alpha\beta M} = (\pi_{y1M}\gamma_z^{\,0})_{\alpha\beta} = (\pi_{z1M}\gamma_y^{\,0})_{\alpha\beta} = \psi_2{}^{\dagger}{}_{\alpha\beta M} \tag{5.11}$$
$$\pi_{2\alpha\beta M} = (\pi_{y2M}\gamma_z^{\,0})_{\alpha\beta} = (\pi_{z2M}\gamma_y^{\,0})_{\alpha\beta} = \psi_1{}^{\dagger}{}_{\alpha\beta M} \tag{5.12}$$

The form of the conjugate momenta implies that the only non-zero equal time anticommutators are:[39]

$$\{\pi_{j\alpha\beta M}(\mathbf{y}, y^0, \mathbf{z}, z^0),\ \psi_{i\kappa\lambda M}(\mathbf{y'}, y^0, \mathbf{z'}, z^0)\} = \{\ \psi_{j\alpha\beta M}{}^{\dagger}(\mathbf{y}, y^0, \mathbf{z}, z^0),\ \psi_{i\kappa\lambda M}(\mathbf{y'}, y^0, \mathbf{z'}, z^0)\}$$
$$= (1 - \delta_{ij})\,\delta_{\alpha\kappa}\,{}_{\beta}\delta_{\beta\lambda}\,\delta^{r-1}(\mathbf{y} - \mathbf{y'})\delta^{r'-1}(z - \mathbf{z'}) \tag{5.13}$$

for i, j = 1, 2.

If we introduce fractionation s where the particle is fractionated to s gambols, and introduce an internal symmetry index γ, then the free Megaverse PseudoFermion wave function has the form:

$$\psi_{i\alpha\beta\gamma M}(y,z) = \sum_{s}\sum_{s_1}\sum_{s_2}\int dp^{r'-1}\int dq^{r-1} N(p, q)[b_{\gamma iM}(s, p, q, s_1, s_2)u_\alpha(p,s_1)u_\beta(q,s_2)\exp(-ip\cdot y - iq\cdot z) +$$
$$+\ d_{\gamma iM}(s, p, q, s_1, s_2)^{\dagger}v_\alpha(p,s_1)v_\beta(q,s_2)\exp(ip\cdot y + iq\cdot z)] \tag{5.14}$$

plus Hermitean conjugates for i = 1, 2 where N(p, q) is a normalization factor.

The composite creation and annihilation operators satisfy the anti-commutation relations:

$$\{b_{\gamma iM}(s,p,q,s_1,s_2),\ b_{\gamma' jM}(s',p',q',s_1',s_2')^{\dagger}\} = \{d_{\gamma iM}(s, p, q, s_1, s_2),\ d_{\gamma' jM}(s', p', q', s_1', s_2')^{\dagger}\} =$$
$$= (1 - \delta_{ij})\,\delta_{\gamma\gamma'}\delta_{s,s'}\delta_{s_1,s_1'}\delta_{s_2,s_2'}\delta^{r-1}(\mathbf{p} - \mathbf{p'})\delta^{r'-1}(\mathbf{q} - \mathbf{q'})\,U_g(\varepsilon(s,q)) \tag{5.15}$$

where U_g is given by eqs. 2. 22.

The other composite anti-commutation operators are zeroes.

$$\{b_{\gamma iM}(s, p, q, s_1, s_2),\ b_{\gamma' jM}(s', p', q', s_1', s_2')\} = 0 \tag{5.16}$$
$$\{b_{\gamma iM}(s, p, q, s_1, s_2)^{\dagger},\ b_{\gamma' jM}(s', p', q', s_1', s_2')^{\dagger}\} = 0$$
$$\{d_{\gamma iM}(s, p, q, s_1, s_2),\ d_{\gamma' jM}(s', p', q', s_1', s_2')\} = 0$$
$$\{d_{\gamma iM}(s, p, q, s_1, s_2)^{\dagger},\ d_{\gamma' jM}(s', p', q', s_1', s_2')^{\dagger}\} = 0$$

and anti-commutators of b and d type operators are zero as well.

[39] See S. Blaha, Il Nuovo Cimento **49A**, 35 (1979) for one coordinate system, PseudoQuantum fermions.

5.2 Universe Gambol Operators

The creation/annihilation operators may be divided into Megaverse particle and universe gambol operators as in sections 3.2 – 3.5 where the particle operators become Megaverse particle creation and annihilation operators while the form of gambol operators are universe gambol operators. *The only change is to view the universe gambols within four space-time dimension subspaces of the six space-time dimension Megaverse. Thus we effectively set r' = r below*[40].

The internal symmetries of the Megaverse are a quadruple of those in our universe. Thus we must also restrict the set of Megaverse symmetries to those of our universe. We specify a γ index for the universe internal symmetries.[41]

The spin of a Megaverse fermion is different from that of a fermion in the universe. We restrict the spin of the fermions in the four dimension Megaverse subspace to that of our universe.

5.3 Universe Gambol Creation/Annihilation Operators

We define the universe gambol operators with

$$\begin{aligned} b^{1/s}{}_{g\gamma iM}(s, q, s_2) &= \Sigma_{s_1} (\int d^{r-1}pN'(p,q))^{\frac{1}{2}}\, b_{\gamma iM}(s, p, q, s_1, s_2) \\ b^{1/s}{}_{g\gamma iM}(s, q, s_2)^{\dagger} &= \Sigma_{s_1} (\int d^{r-1}p\, N'(p,q))^{\frac{1}{2}}\, b_{\gamma iM}(s, p, q, s_1, s_2)^{\dagger} \end{aligned} \tag{5.17}$$

where s is the fractionation. Then we define the gambol universe field operator anti-commutators with the anti-commutation relations:

$$\begin{aligned} &\{b^{1/s}{}_{g\gamma iM}(s, q, s_2), b^{1/s'}{}_{g\gamma' jM}(s', q', s_2')^{\dagger}\} = \\ &= \Sigma_{s_1}\Sigma_{s_1'}(\int d^{r-1}pN'(p,q))^{\frac{1}{2}}(\int d^{r-1}p'N'(p',q))^{\frac{1}{2}}\, \{b_{\gamma iM}(s,p,q,s_1,s_2), b'_{\gamma' jM}(s',p',q',s_1',s_2')^{\dagger}\} \\ &= (\int d^{r-1}pN'(p,q))^{\frac{1}{2}}(\int d^{r-1}p'N'(p',q))^{\frac{1}{2}}(1 - \delta_{ij})\, \delta_{s,s'}\, \delta_{s_2 s_2'}\, \delta_{\gamma,\gamma'}\delta^{r-1}(\mathbf{p} - \mathbf{p'})\delta^{r-1}(\mathbf{q} - \mathbf{q'})\, U_g(\varepsilon(s,q)) \\ &= \int d^{r-1}p\, N'(p,q)\, (1 - \delta_{ij})\, \delta_{s,s'}\, \delta_{s_2 s_2'}\, \delta_{\gamma,\gamma'}\delta^{r-1}(\mathbf{p} - \mathbf{p'})\delta^{r-1}(\mathbf{q} - \mathbf{q'})\, U_g(\varepsilon(s,q)) \\ &= \int d^{r-1}p\, N'(p,q)\, \delta^{r-1}(\mathbf{p} - \mathbf{p'})\, (1 - \delta_{ij})\, \delta_{s,s'}\, \delta_{s_2 s_2'}\, \delta_{\gamma,\gamma'}\delta^{r-1}(\mathbf{q} - \mathbf{q'})\, U_g(\varepsilon(s,q)) \\ &= (1 - \delta_{ij})\, \delta_{s,s'}\delta_{s_2 s_2'}\, \delta_{\gamma,\gamma'}\delta^{r-1}(\mathbf{q} - \mathbf{q'})\, U_g(\varepsilon(s,q)) \end{aligned} \tag{5.18}$$

and

$$\{d^{1/s}{}_{g\gamma iM}(s, q, s_2), d^{1/s'}{}_{g\gamma' jM}(s', q', s_2')^{\dagger}\} = (1 - \delta_{ij})\, \delta_{s,s'}\delta_{s_2 s_2'}\, \delta_{\gamma\gamma'}\delta^{r-1}(\mathbf{q} - \mathbf{q'})\, U_g(\varepsilon(s,q)) \tag{5.19}$$

where eqs. 2.22 specifies $U_g(\varepsilon(s,q))$ with m = M.

Eqs. 3.18 and 3.19 have fractional integrations that utilize

$$[(\int dp^{r-1}N'(p,q))^{\frac{1}{2}}]^2 = \int dp^{r-1}\, N'(p,q) \tag{5.20}$$

---

[40] The purpose of this choice is to facilitate the map between Megaverse and universe operators in section 5.5.

[41] In chapter 10 we show the Cosmos PseudoFermion formulation automatically gives restricted Internal Symmetries to the gambol quantum fields. The number of γ index values is d_{dN} where N is the Blaha Number of the universe's space.

as in Riemann-Liouville integrals. Note γ and γ' are universe internal symmetry indices.

Thus the Megaverse subspace and the universe gambols have the same internal symmetries and spin. They differ in mass and momentum. The corresponding d type operators have similar anti-commutation relations. The other anti-commutators are zero:

$$\begin{aligned}
&\{b^{1/s}{}_{g\gamma iM}(s, q, s_2)^\dagger, b^{1/s'}{}_{g\gamma' jM}(s', q', s_2')^\dagger\} = 0 \\
&\{d^{1/s}{}_{g\gamma iM}(s, q, s_2)^\dagger, d^{1/s'}{}_{g\gamma' jM}(s', q', s_2')^\dagger\} = 0 \\
&\{b^{1/s}{}_{g\gamma iM}(s, q, s_2), b^{1/s'}{}_{g\gamma' jM}(s', q', s_2')\} = 0 \\
&\{d^{1/s}{}_{g\gamma iM}(s, q, s_2), d^{1/s'}{}_{g\gamma' jM}(s', q', s_2')\} = 0
\end{aligned} \tag{5.21}$$

and anti-commutators of b and d type operators are zero as well.

The gambol creation/annihilation operators above can be used to define universe gambol fermion quantum fields.

$$\begin{aligned}
\psi_{g\gamma M1}(x) &= \Sigma_p\, [b^{1/s}{}_{g\gamma 1M}(p, s)\, f_p(x) + d^{1/s}{}_{g\gamma 1M}(p, s)^\dagger f_p{}^*(x)] \qquad (5.22)\\
\psi_{g\gamma M2}(x) &= \Sigma_p\, [b^{1/s}{}_{g\gamma 2M}(p, s) f_p(x) + d^{1/s}{}_{g\gamma 2M}(p, s)^\dagger\, f_p{}^*(x)]
\end{aligned}$$

5.4 Megaverse Particle Creation/Annihilation Operators

The Megaverse particle operators may be defined using the composite creation/annihilation operators of eq. 5.14 for particles of momentum p. We replace[42] the sum[43] over s with a sum over energies ε so as to take advantage of the normalization sum in eq. 2.21:[44]

$$b_{\gamma iM}(p, s_1) = \int_0^\infty d\varepsilon\, \Sigma_{s_2}\, (\int d^{r-1}q\, N'(p,q))^{1/2}\, b_{\gamma iM}(s, p, q, s_1, s_2) \tag{5.23}$$

$$b_{\gamma iM}(p, s_1)^\dagger = \int_0^\infty d\varepsilon\, \Sigma_{s_2}\, (\int d^{r-1}q\, N'(p,q))^{1/2}\, b_{\gamma iM}(s, p, q, s_1, s_2)^\dagger$$

The anti-commutation relations are[45,46]

$$\begin{aligned}
&\{b_{\gamma iM}(p, s_2), b_{\gamma' jM}(p', s_2')^\dagger\} = \\
&\quad = \int d\varepsilon \int d\varepsilon'\, \Sigma_{s_1}\Sigma_{s_1'} (\int d^{r-1}q N'(p,q))^{1/2} (\int d^{r-1}q' N'(p,q'))^{1/2} \{b_{\gamma iM}(s,p,q,s_1,s_2), b_{\gamma jM}{}^\dagger(s',p',q',s_1', s_2')\} \\
&\quad = \int d\varepsilon \int d\varepsilon' \Sigma_{s_1}\Sigma_{s_1'} (\int d^{r-1}q N'(p,q))^{1/2} (\int d^{r-1}q' N'(p,q'))^{1/2} (1 - \delta_{ij})\, \delta_{s_1,s_1'}\, \delta_{s_2,s_2'} \delta_{ss'} \delta_{\gamma\gamma'} \delta^{r-1}(\mathbf{p} - \mathbf{p'}) \cdot \\
&\qquad \cdot \delta^{r-1}(\mathbf{q} - \mathbf{q'}) U_g(\varepsilon(s,q)) \\
&= \int d^{r-1}q\ \ N'(p,q)\, (1 - \delta_{ij})\, \delta_{ss'} \delta_{s_2,s_2'}\, \delta_{\gamma\gamma'} \delta^{r-1}(\mathbf{p} - \mathbf{p'}) \delta^{r-1}(\mathbf{q} - \mathbf{q'}) \int d\varepsilon\ U_g(\varepsilon(s,q)) \\
&= \int d^{r-1}q\, N'(p,q)\, \delta^{r-1}(\mathbf{q} - \mathbf{q'})\, (1 - \delta_{ij})\, \delta_{ss'} \delta_{s_2,s_2'}\, \delta_{\gamma\gamma'} \delta^{r-1}(\mathbf{p} - \mathbf{p'}) \\
&= (1 - \delta_{ij})\, \delta_{ss'} \delta_{s_2,s_2'}\, \delta_{\gamma\gamma'} \delta^{r-1}(\mathbf{p} - \mathbf{p'})
\end{aligned} \tag{5.24}$$

---

[42] We use S and ε interchangeably.

[43] The sum over S begins as a discrete sum over powers of 2 according to Limos. We replace it with a continuous value for S, which makes the transition to an integral over ε possible.

[44] Using a sum over ε supports eq. 5.13.

[45] The S, S' and $\delta_{SS'}$ terms are equivalent to ε energy terms due to the 1:1 relation of S and ε specified in eq. 2.23.

[46] We designate the spin as s_2 for later use in section 5.5.

using eq. 3.21 with ε(s) and ε(s'). We make the s (and ε) sums into integrations for analytic convenience just as the discrete sum over hν in the black body Planck distribution derivation is similarly made continuous.

The other anti-commutation relations are:

$$\{d_{\gamma iM}(p, s_2), d_{\gamma' jM}(p', s_2')^\dagger\} = (1 - \delta_{ij})\, \delta_{ss'}\delta_{s_2,s_2'}\delta_{\gamma\gamma'}\delta^{r-1}(\mathbf{p} - \mathbf{p}') \quad (5.25)$$
$$\{b_{\gamma iM}(p, s_2), b_{\gamma' jM}(p', s_2')\} = 0$$
$$\{d_{\gamma iM}(p, s_2), d_{\gamma' jM}(p', s_2')\} = 0$$
$$\{b_{\gamma iM}(p, s_2)^\dagger, b_{\gamma' jM}(p', s_2')^\dagger\} = 0$$
$$\{d_{\gamma iM}(p, s_2)^\dagger, d_{\gamma' jM}(p', s_2')^\dagger\} = 0$$

and anti-commutators of b and d type operators are zero as well.

The creation/annihilation operators above are those that appear in Megaverse fermion quantum fields. Thus we may define a Megaverse particle quantum field with

$$\psi_{\gamma M1}(x) = \Sigma_p\, [b_{\gamma 1M}(p, s)\, f_p(x) + d_{\gamma 1M}(p, s)^\dagger f_p{}^*(x)] \quad (5.26)$$
$$\psi_{\gamma M2}(x) = \Sigma_p\, [b_{\gamma 2M}(p, s) f_p(x) + d_{\gamma 2M}(p, s)^\dagger\, f_p{}^*(x)]$$

with internal symmetry index γ.

5.5 Map Between Universe Gambol and Megaverse Particle States

We now relate the Megaverse particle wave function to sets of gambol wave functions by requiring a symmetry in the creation/annihilation operators:

$$b_{\gamma iM}(s, p, q, s_1, s_2) = b_{\gamma iM}(s, q, p, s_2, s_1) \quad (5.27)$$
$$d_{\gamma iM}(s, p, q, s_1, s_2) = d_{\gamma iM}(s, q, p, s_2, s_1)$$

plus Hermitean conjugates. Then we may relate the gambol operators to the particle operators using eqs. 3.17 and 3.22:

$$b_{\gamma iM}(p, s_2) = \int d\varepsilon \Sigma_{s_1} (\int d^{r-1}q\, N'(p,q))^{1/2}\, b_{\gamma iM}(s, p, q, s_1, s_2) \quad (5.28)$$
$$= \int d\varepsilon\, b^{1/s}{}_{g\gamma iM}(s(\varepsilon), p, s_2)$$

Thus the Megaverse particle operator is a sum of universe gambol operators. *The universe gambols have the same spin, momentum, and internal symmetry as the parent Megaverse particle.*[47] The gambol mass, m_g = m/s, appears in ε(s,p). The gambol temperature is defined in terms of the particle mass: kT_g = 0.0785M. (See Blaha (2023e).)

These features may be seen in the form of $U_g(\varepsilon(s, p))$ in eqs. 2.5 and 2.22. Each term in eq. 5.28 defines an operator for a set of gambols from which one is projected in

---

[4747] Due to the considerations of section 5.2.

a Megaverse particle interaction. The particle interacts via a sum of gambol $b^{1/s}{}_{g\alpha iM}(s, p, s_2)$ gambol interactions.

5.6 Megaverse Particle – Universe Gambol States

A Megaverse one particle state of fractionation is

$$|1\rangle = b_{g\gamma 2M}(s, p, s_2)^{\dagger}|0\rangle \quad (5.29)$$

It is equivalent to a sum of sets of gambols for all values of s Its bra is

$$\langle g| = \langle 0|\Sigma_s\, b^{1/s}{}_{g\gamma 1M}(s, p, s_2)^{\dagger} = \langle 1| \quad (5.30)$$

by eq. 3.27. (Note PseudoQuantum formulation.)

If we project a universe gambol from a Megaverse particle we find the probability distribution

$$\langle 1|g\rangle = \langle g\ s, p, s_2|g\rangle = U_g(\varepsilon(s, p)) \quad (5.31)$$

We now have a quantum field theory for a Megaverse particle – universe gambol configuration.

5.7 Individual Universe Gambols

The above formalism projects a universe gambol from a set of gambols of fractionation s. The individual gambols in a set have identical quantum numbers. But the sets are defined as conceptual assemblages of gambols due to a fractionation of a Megaverse particle within a Megaverse subspace into sets of gambols. A gambol does not exist until it is projected from an assemblage by a projection operator[48] $b^{1/s}{}_{g\alpha 2}(s, p, s_2)^{\dagger}$. Then it may interact.

In this model our universe would appear as a gambol generated probabilistically in the Megaverse.

5.8 Multiple Universes

Multiple universes may be defined in a Megaverse. A multiple universe quantum field state can be expressed in terms of k universe gambol creation operators:

$$|g\ k\rangle = b^{1/s_1}{}_{g\gamma 2M}(s_1, p_1, S_1)^{\dagger}\, b^{1/s_2}{}_{g\gamma 2M}(s_2, p_2, S_2)^{\dagger}|0\rangle \ldots b^{1/s_k}{}_{g\gamma 2M}(s_k, p_k, S_k)^{\dagger}|0\rangle \quad (5.32)$$

\

where S_i is a spin for i = 1, 2, … k. The fractionations s_i may or may not be equal. The fractionations, momentum spin combinations (s_i, p_i, S_i) must be distinguishable due to eq. 2.9. Thus a Megaverse may support multiple universes.

48 The subscript 2 is due to our PseudoQuantum Theory of fields.

6. Gamboled Hubble Expanding Universe

We have seen that a Planckian[49] probability distribution applies to particles in Blaha (2023e).[50] We now consider the case of the mass-energy of a dynamic confined universe treated as an expanding black body with a decreasing temperature.

We begin by reconsidering the Einstein derivation of the Planck distribution for the case of a slowly changing (adiabatic) black body of increasing size and decreasing temperature.

6.1 Universe Particle Gambols

We have previously developed the possibility that universes are expanding particles in previous books: *Universes are Particles* (2021), *Newton's Apple is Now the Fermion* (2023) and earlier books. We suggest that our universe (and others) were created in a parent, higher dimension universe, the Megaverse, as a "point" object. We further suggest a new universe is an unstable object describable as a set of gambols. It expands in a similar way to the gambol decay/expansion of a particle resonance into decay products.

In view of our Gambol Model of particles it is reasonable to suggest that a universe begins life as a point of great mass-energy with a gambol probability distribution. The sets of gambols in this novo-universe offer many possibilities for its expansion. After beginning expansion, the set of gambols reconstitutes and proceeds to undergo Hubble-like expansion.[51] We thus view the expansion of the universe, parameterized a fractionation into time dependent parts, as a process in time. We set the fractionation as $s = 1/bt$ where b is a constant and t is the time. We then extend the discrete set of fractionations into 1/s parts to be continuous using the continuous variable t.

We see the universe starting out as an infinite, uniform set of infinitesimal gambol parts as growing to a small set of gambols as time progresses with the gambols "specifying" the universe's clumpiness in density.

6.2 Einstein Derivation of Planck Distribution for a Fixed Black Body

Einstein derived the Planck distribution following a statistical approach. Einstein's[52] derivation starts with the concept of quantum jumps between energy states $E_1 < E_2$ due to a radiation density u.

---

[49] This distribution was first developed for the case of photons confined to a black body of fixed size.

[50] If the black body should expand gradually one would see a decreasing temperature and varying Planck distribution that becomes broader. This situation parallels the Hubble expansion of the universe with its decreasing temperature seen in this chapter.

[51] At $t = 0$, the fractionation is $s = \infty$, and the universe has ultimate fractionation with zero clumpiness of the univere's mass-energy. As time progresses, clumpiness increases as astrophysical experiment suggests. Later we use this concept to derive a relation between the S_8 Clumpiness Index and universe fractionation.

[52] A. Einstein, Verhandl, deut. Physic. Ges. **18**, 318 (1916); Physik. Z. **18**, 121 (1917); J. W. M. Dumond and E. R. Cohen, Revs. Mod. Physics **25**, 691 (1953).

There are three ways a quantum jump occurs: 1. spontaneous jumps with probability S_{12} for a transition from E_2 to E_1 with a quantum emission; 2. absorption of a quantum giving a jump from E_1 to E_2 with probability $A_{21}u$; and 3. induced emission of a quantum giving a jump from E_1 to E_2 with probability $I_{12}u$. The probabilities that states 1 and 2 are occupied in thermal equilibrium are proportional to $\exp(-E_1/kT)$ and $\exp(-E_2/kT)$ respectively. Thus

$$(S_{12} + I_{12}u) \exp(-E_2/kT) = A_{21}u \exp(-E_1/kT) \qquad (6.1)$$

expresses the balance of transitions. The constants S_{12}, I_{12} and A_{21} are properties of the states and independent of the density u. At high temperature $S_{12} << I_{12}u$. This implies

$$I_{12} = A_{21} \qquad (6.2)$$

The transitions: $1 \rightarrow 2$ and $2 \rightarrow 1$ induced by the electromagnetic field are the same.

The density u may be calculated to be

$$u = (S_{12}/I_{12})/(\exp((E_2 - E_1)/kT) - 1) \qquad (6.3)$$

Einstein's derivation assumes that the container and the radiation can coexist in a stable state and that transitions between states can be determined statistically.

6.3 Our Derivation of an Anti-Planckian Distribution for a Growing Black Body

We see the distribution needed for the Hubble growth of our universe requires a different form.[53] Our derivation also starts with the concept of quantum universe jumps between energy states $E_1 < E_2$ of a universe due to a radiation density w in states (stages) of different size if one wishes to develop a distribution for universe expansion. Initially the states can be viewed as discrete. Then they can be taken to be continuous.

We now consider the case where the average size of the universe changes with time. We assume the possible states for the average size are initially discrete and undergo "quantum" jumps. Then we progress to the continuous case.

There are three ways a quantum jump occurs:

[53] This derivation differs from that of chapter 4 since it has different physical characteristics. We view gambols as a theoretical construct and not as real entities. An entity may be treated in different ways with different gambol models with each suited to a specific purpose. (Chapter 10.) Each gambol fractionation is theoretical. So one cannot say that there is one and only one specific fractionation into a specific set of "real" gambols. The gambol concept is thus more abstract than the quark concept since quarks, although confined, are real – meaning we can directly measure them as particles. Gambols adopt the characteristics of their parent particle. Gambols appear in sets of identical gambols. Creation operators project out an individual gambol to represent a particle in an interaction. The possibility of constructs like the gambol construct where there is a precursor entity that is not directly experimentally measureable but leads to experimentally measurable results was considered by 20th Century Philosophers of Science such as R. B. Braithwaite (*Scientific Explanation*, Cambridge University Press, 1953) and Karl Popper whose thesis was that Physical theories can never be proven but can be falsified and shown to be true or false via experiment.

- Spontaneous with probability wS_{12} for a transition from E_2 to E_1 with a quantum emission where w is the energy density;

- With probability A_{21} absorption of a quantum giving a jump from E_1 to E_2,

- With probability I_{12} for induced emission of a quantum giving a jump from E_1 to E_2

The S_{12} transition appears with a w energy density factor since it is dependent on the available energy density: the higher the density the more favored the jump. *In the case of an expanding universe spontaneous jumps are proportional to the mass-energy (radiation) density w. Absorption and induced emission are not proportional to w in this expanding universe.*

The probabilities that universe states 1 and 2 are occupied in thermal equilibrium are proportional to $\exp(-E_1/kT)$ and $\exp(-E_2/kT)$ respectively implying

$$(S_{12}\, w + I_{12}) \exp(-E_2/kT) = A_{21} \exp(-E_1/kT) \tag{6.4}$$

which expresses the balance of transitions. Thus w = 1/u if we compare to eq. 6.1 We call the w energy density *anti-Planckian*. The constants S_{12}, I_{12} and A_{21} are properties of the states and independent of the density w. At high universe temperature $S_{12}v >> I_{12}$ and A_{21}. We may thus approximate

$$I_{12} = A_{21} \tag{6.5}$$

Implementing the equality of the induced transitions: $1 \rightarrow 2$ and $2 \rightarrow 1$.

The density w is then

$$w = (I_{12}/S_{12})(\exp((E_2 - E_1)/kT) - 1)$$

$$w = N\,(\exp(\varepsilon/kT) - 1) \tag{6.6}$$

where N is a normalization constant embodying (I_{12}/S_{12}) and $\varepsilon = E_2 - E_1$. We assume with Einstein that the container (the universe boundary[54]) and the energy density can coexist in a quasi-stable (adiabatic) state and that transitions between states can be determined statistically.

We now introduce the number of nodes in the interval $(\varepsilon, \varepsilon + d\varepsilon)$

$$8\pi\varepsilon^2 \tag{6.7}$$

giving the distribution (using the gambol temperature T_g)

---

[54] See chapter 7 for the derivation of Hubble expansion partly based on Casimir force confinement at the universe boundary.

$$W(\varepsilon(t), t) = 8\pi N\, \varepsilon^2 (\exp(\varepsilon/kT_g) - 1) = H(t) \tag{6.8}$$

which we will show below is the Hubble parameter.

6.4 Time and the Hubble Parameter

The Planck distribution is a time independent distribution for the energy distribution of photons in a black body. In Blaha (2023e) we extended the role of the Planck distribution to a Planckian probability distribution of the gambol energy of a particle.

We now raise the possibility that the anti-Planckian distribution derived above has the role of describing the growth of the universe parameterized by the Hubble parameter. The previous uses of Planck-type distributions were for black bodies (of fixed time-independent size) and for gamboled particles that were time independent.

We suggest that a universe particle[55] has a changing time dependent Hubble distribution. We view the universe as a form of resonant particle, which decays (expansion) in time, that is created at one instant in a Megaverse and proceeds to decay ultimately to a steady state of fixed size. The anti-Planckian distribution measures the rate of expansion (decay) as a function of time. The declining time dependence of the universe temperature, which appears in $W(\varepsilon)$ in the form of kT supports the time dependent Hubble interpretation.

The Planck distribution specifies the number of photons of a specific energy. We have shown that it can be viewed as a probability distribution in an experiment where one makes a pin hole in a black body and counts emerging photons. They must be viewed in accord with the Planck distribution treated as a probability distribution. This view leads to our Planckian probability distribution for particles.

In the case of a universe, the changes in state as a universe expands may be viewed as a measurable quantity. These changes are presently embodied in the Hubble parameter. We have shown they are the result of spontaneous jumps, with which we can associate a probability that depends on the mass-energy ε and temperature kT of the universe state. The probability then becomes a measure of the change. We equate the anti-Planckian probability, suitably normalized, to the Hubble parameter, which is a measure of the change of the universe state.

We develop the universe Anti-Planckian Model of our universe below. Other universes may be expected to have a similar model.

6.5 Calculation of the Gambol-based Hubble Parameter

The Gambol Model is based on a fractionation parameterized by a constant s, which is the number of parts into which an entity is split by a fractionation.[56] The size of each part is 1/s of the entity.

---

[55] See Blaha (2021d) *Universes are Particle* and earlier work.

[56] This somewhat sensitive calculation is performed in double precision using Excel. The calculations in Blaha (2023e) are also performed in double precision using Excel. Double precision mathematics was also required in the author's calculation of the Fine Structure Constant since it is known to thirteen decimal places. That calculation found the exact known result.

Turning to the inputs for the calculation: we bring in time dependence by defining a constant b with our empirical value (needed to implement the Gambol Hubble Model):

$$b = \; = 8.67 \times 10^{-19}\ \text{sec}^{-1} \tag{6.9}$$

for use in transforming s into an equivalent time

$$s = 1/(bt) \tag{6.10}$$

We define the gambol energy parameter with

$$\varepsilon(s) = [(M_g\, s/M + 1)/(s + 1)]\ E \tag{6.11a}$$

or, in terms of time,

$$\varepsilon(t) = [(M_g\, /M\ + bt)/(bt)]\ E \tag{6.11b}$$

where M is the total mass-energy of the universe, M_g is the mass of a universe gambol, and E = M is the total mass-energy of the universe. The choice of E = M would indicate the universe is not "moving." Note: we omit a "1" in forming eq. 6.11b so that the Hubble parameter will be infinite at t = 0 – the Big Bang.

Eq. 6.11 is intentionally analogous to the definition of the gambol parameter for particles in Blaha (2023e).

We define the *gambol* temperature T_g of the universe with

$$kT_g = 0.0785\ M\ s\ = 0.0785\ M/(bt) \tag{6.12}$$

where M is the mass of the universe using the same definition as for particles in Blaha (2023e). Note: kT_g is infinite at t = 0 – the Big Bang.

6.5.1 Determination of Constants[57]

The **total** mass-energy[58] of the universe has been *estimated* to be

$$M = 1.712 \times 10^{81}\ \text{GeV/c}^2\ = E \tag{6.13}$$

and the universe gambol mass is

$$M_g = M/1024 \tag{6.14}$$

The age of the universe has been estimated to be

$$t_{NOW} =\ 13.787 \text{ billion years} = 4.35 \times 10^{17}\ \text{sec} \tag{6.15}$$

The currently known Hubble parameters are

---

[57] The extremely large size of some variables and the extremely small size of other variables causes the calculation to be sensitive to the parameters.

[58] This value includes normal mass and energy plus Dark Energy and Dark Matter.

$$H_0 = \text{Hubble rate now} = 67.4 \text{ km/s}^{-1} \text{ Mpc}^{-1} \tag{6.16}$$
$$\text{Hubble length} = c/H_0 = 1.372 \times 10^{26} \text{ m}$$
$$H_0 = 3*10^7/1.372 \times 10^{26} \text{ sec}^{-1} = 2.19 \times 10^{-19} \text{ sec}^{-1}$$
$$1/H_0 = 4.57 \times 10^{18} \text{ sec}$$

The known values of the Hubble parameter are

$$\text{At } t = 380000 \text{ years} = 1.2 \times 10^{13} \text{ sec} \quad H = 67.8 \text{ km s}^{-1} \text{ Mpc}^{-1}$$
$$\text{At } t = t_{NOW} \quad H = 73.2 \text{ km s}^{-1} \text{ Mpc}^{-1} \tag{6.17}$$

The universe temperature now is

$$kT_{NOW} = k\, 2.72548 \text{ K} = 2.126 \times 10^8 \text{ GeV/c}^2 \tag{6.18}$$

The present physical time T is related to the present T_g by

$$kT_{NOWphys} = 5.97 \times 10^{-73}\, kT_{gNOW} \tag{6.19}$$

The anti-Planckian normalization in eq. 6.9 is

$$N = 8.078 \times 10^{-165} \text{ km s}^{-1} \text{ Mpc}^{-1} \text{ - s}^2 \tag{6.20}$$

6.6 The Gambol Hubble Model Results

The Gambol Hubble Model generates an Anti-Planckian distribution that corresponds to the known Hubble features. See Figs. 6.1 - 6.3. The values and their times (eq. 6.17) are successfully[59] obtained. In addition,

1. The calculation involves both extremely large and extremely small quantities. It is constrained by the framework of gamboled particle equations and parameter values found in *particle* (quark, lepton and hadron) calculations. It gives a convincing formulation of the time dependence of the Hubble parameter by treating universes as particles whose expansion/evolution mirrors particle decay.

2. The t = 0 "infinite" Hubble parameter value is obtained.

3. The distribution has a "Big Dip", similar to the author's empirical Hubble fit appearing in earlier books. Simple algebra and calculus shows that a latter value (72.3) larger than a smaller value (67.8) necessitates a minimum in the Hubble parameter between these points.

[59] Should the experimental Hubble parameter values change, then the parameters of the calculation, including the M_g fractionation s, b, and E can be changed appropriately.

4. There is an interregnum, we have called the "Big Dip", (Fig. 6.1) between t = 10^{14} and 10^{17} where the Hubble parameter has very small values ranging between 0.051 and 1.27. This region suggests a slower period of universe expansion.

5. The rise after the interregnum may constitute a Big Bang "Second Expansion" phase. This phase does not appear to signal new Physics since it is embodied in the Anti-Planckian distribution in time.

6. The gambol temperature satisfies an equation similar to the gambol temperature equation in Blaha (2023e) for quarks, leptons, and hadrons:

$$kT_{gu} = 0.0785\ M_u\ s\ = 0.0785\ M_u/(bt) \tag{6.12}$$

with the addition of a time dependence since the evolution of the universe is for a long period of time whereas the dynamics of interactions in Blaha (2023e) is effectively instantaneous. There is a similarity to the Hawking black-body temperature outside black hole event horizon

$$kT = 1/(8\pi GM_{Hawkingu}) \qquad \text{Hawking Mass}$$

suggesting our universe may be a form of Black Hole. The relation of the two temperatures is summarized (neglecting the factor s) by

$$M_u \equiv 1/(0.0875 \times 8\pi GM_{Hawkingu})$$

From the value of $M_u = 1.712 \times 10^{81}$ GeV/c^2 we find the equivalent Hawking mass using

$$M_{Hawkingu} \equiv 1/(0.0875 \times 8\pi GM_u)$$

to find

$$M_{Hawkingu} = 4 \times 10^{-44}\ \text{GeV/c}^2$$

with a corresponding event horizon.

The black hole temperature exists outside the event horizon. It would be interesting to see if there is a black hole temperature just inside the event horizon that corresponds to the gambol temperature kT_{gu} within the universe.

7. The scale factor, usually denoted a, may be determined in this model from[60]

$$\varepsilon(t) = [(M_g/M\ + bt)/(bt)]\ E \tag{6.11}$$

$$W(\varepsilon) = 8\pi N\ \varepsilon^2\ (\exp(\varepsilon/kT_g) - 1) \qquad (.6.8)$$

[60] Note the change in the definition of ε(t) in eq. 6.11. It was necessitated by the required t = 0 Big Bang behavior.

and the relation

$$d/dt \ln (a) = H(t) \tag{6.21}$$

For small t:

$$H(t) \sim 1/t^2$$

implying a Big Bang

$$a(t) \sim \ln(t) \rightarrow -\infty \tag{6.22}$$

For t in the Interregnum region:

$$H(t) \sim \text{small, almost constant vale} = h$$

implying exponential growth

$$a(t) \sim \text{const } e^{ht} \tag{6.23}$$

For t very large in the post-Interregnum region:

$$H(t) \sim \exp(\text{const } t)$$

Using

$$d/dt \ln (a) = H(t)$$

we find "super-exponential" growth

$$\ln (a) \sim \exp(\text{const } t)$$

$$a(t) \sim \exp(\text{const } e^{\text{const } t}) \tag{6.24}$$

Chapter 7 suggests that there might be an end to universe expansion if the decreasing temperature change results in the equality of external Casimir pressure and internal pressure.

8. The gambol factorization for the universe is 1/1024 as opposed to 1/8 for particles. Using s = 1/(bt) we find s = 1024 if t = 1.13×10^{15} sec placing the Hubble minimum at approximately the center of the Interregnum. The anti-Planckian distribution makes the most *probable Hubble point in the distribution* at the Hubble minimum. The Hubble parameter changes with time. The minimum merely indicates the most probable value of s *in the distribution*, namely 1024. *This fact leads to a universe most probable fractionation of 1024.*

The Hubble parameter calculation results in a satisfactory fit to its known values. (A new accurate fit for new Hubble values could be done by mildly adjusting parameters.) It also supports the view that universes are a form of dynamically expanding particle. The next two chapters provide further support for a particle view of universes.

6.7 The Hubble Parameter and Universe Scale Factor

The universe scale factor may be computed from the Hubble parameter using the relation:

$$da/dt\,/a = d/dt \ln a = H(t) = W(t) \tag{6.25}$$

by eq. 6.8. Integrating we find

$$\int_{t_0}^{t} d(\ln a(t')) = \int_{t_0}^{t} da(t')/a(t') = \int_{t_0}^{t} dt'\, H(t') = \int_{t_0}^{t} dt'\, W(\varepsilon(t'), t') \tag{6.26}$$

or

$$a(t) = a(t_0) \exp \left[\int_{t_0}^{t} dt'\, W(\varepsilon(t'), t') \right] \tag{6.27}$$

with $W(\varepsilon(t), t)$ given by eq. 6.8. Eq. 6.27 may be easily evaluated for the t and t_0 cases of small times, Interregnum times and large times. This exercise is left to the reader. Hint: use the approximations in eqs. 6.21 – 6.24.

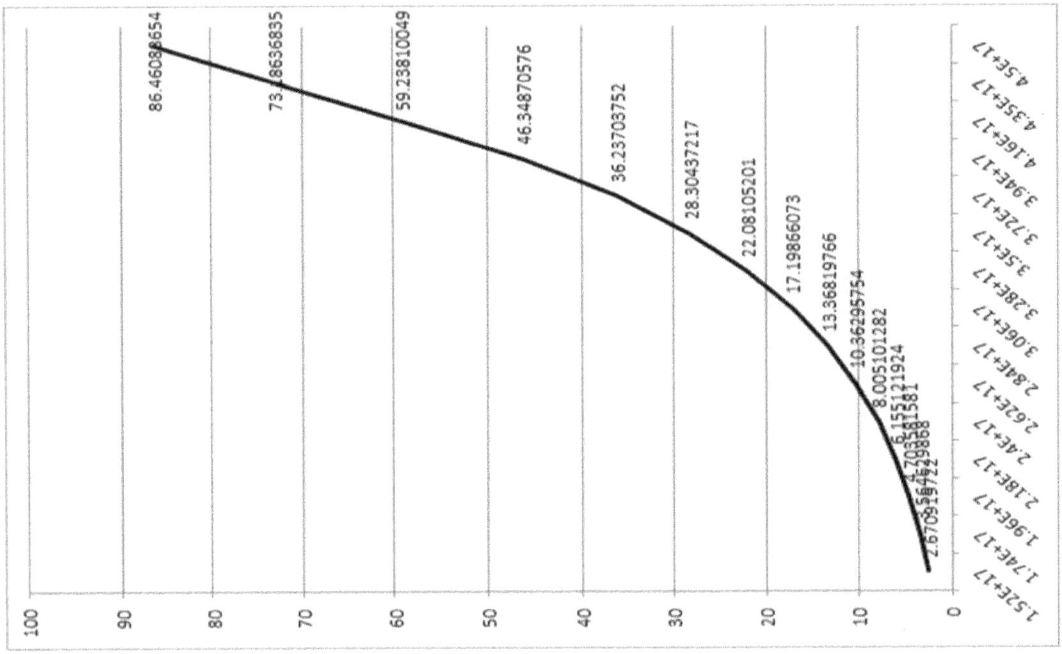

I N T E R R E G N U M   FROM 10^{14} to 10^{17}

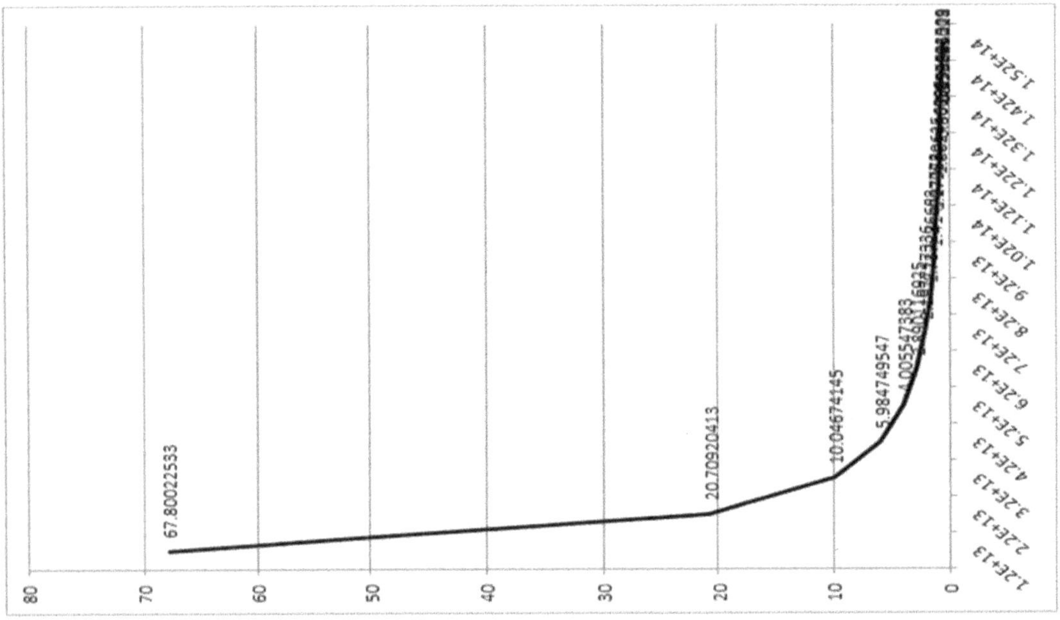

Figure 6.1. The Hubble parameter from 1.3×10^{13} to (now) 4.5×10^{17}.seconds. The vertical axis is the Hubble parameter in km s^{-1} Mpc^{-1} units. The Interregnum is a period of very low Hubble paramter. It is followed by an upswing. The known values of 67.8 and 72.3 km s^{-1} Mpc^{-1} (approximately) are shown.

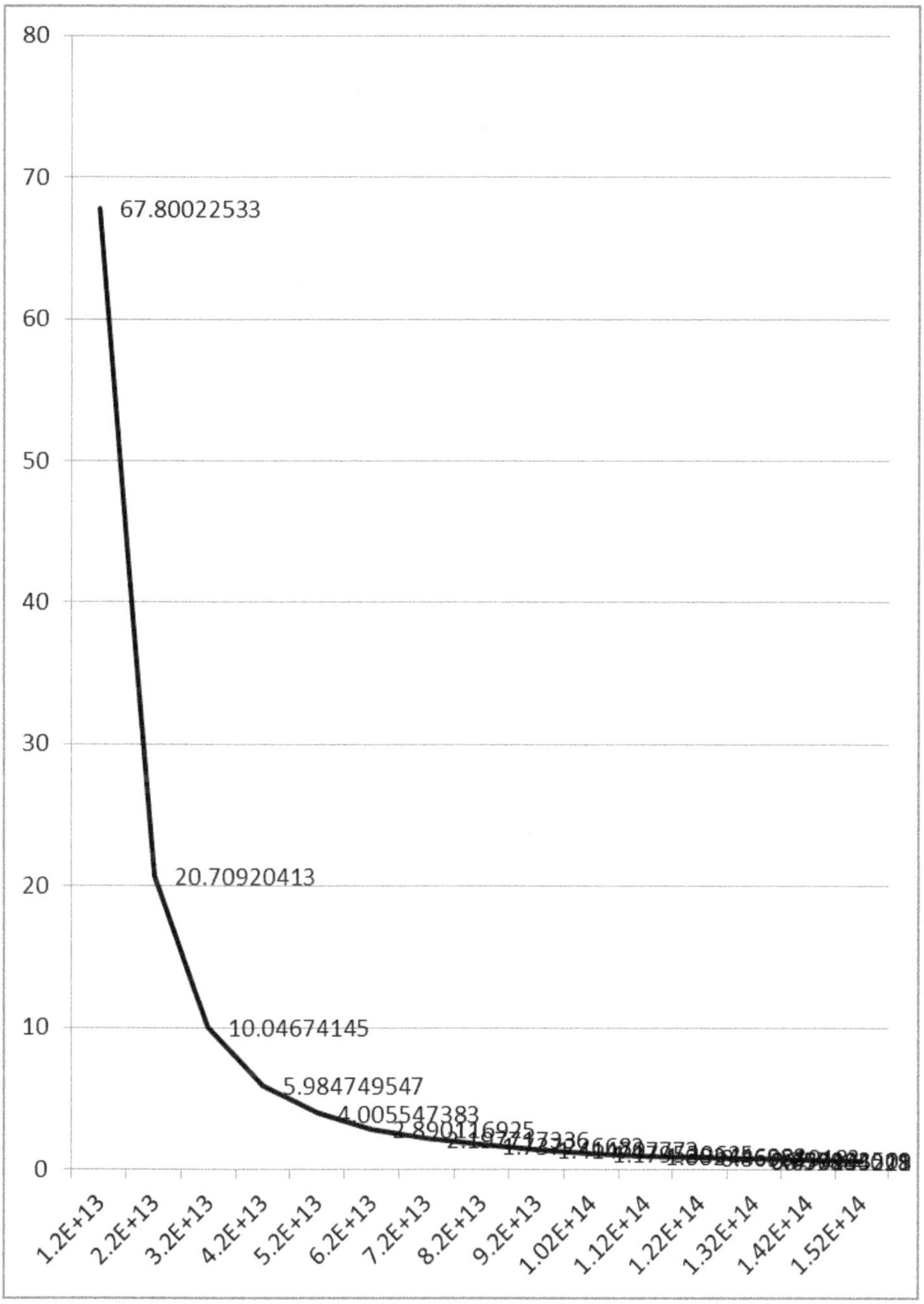

Figure 6.2. The Hubble parameter from 1.2×10^{13} to 1.5×10^{14}.seconds showing the known value of 67.8. The vertical axis is the Hubble parameter in km s^{-1} Mpc^{-1} units.

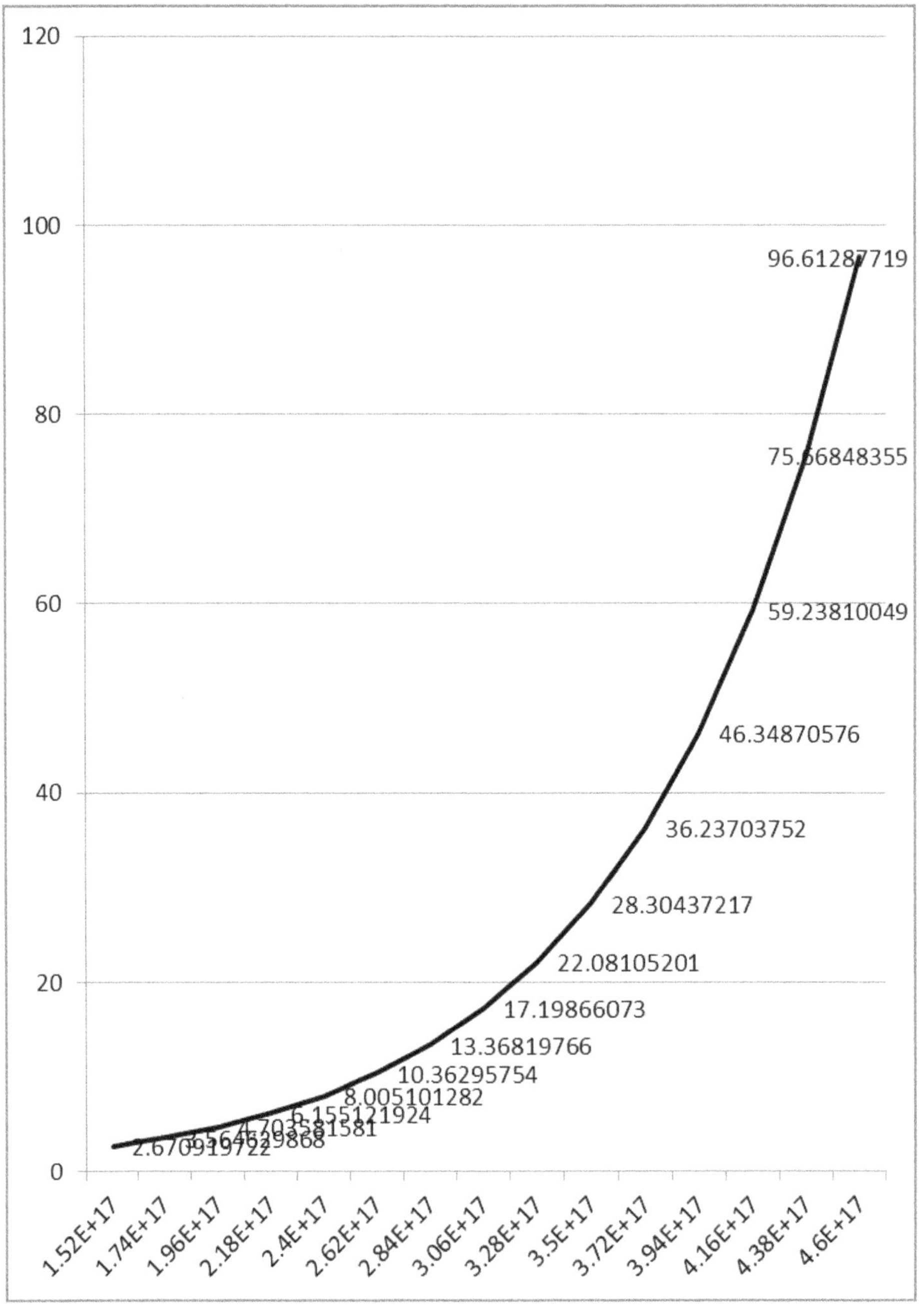

Figure 6.3. The Hubble parameter from 1.52×10^{17} to 4.6×10^{17} seconds containing the known value of 72.3 approximately. The vertical axis is the Hubble parameter in km s^{-1} Mpc^{-1} units.

7. Universe Confinement Through a Vacuum Casimir Effect

In Blaha (2023e) we developed a model for gambol confinement within fundamental particles: quarks and leptons. We now apply the same reasoning to develop a model of universes confined within a larger universe. (Our universe may well be confined within a Megaverse.) We will show that a Casimir force can confine the constituents, matter and energy, of our universe. The universe also has a statistically based outward force due to the pressure of the mass energy within the universe.

The expansion of our universe may be the result of the combined effect of these forces. This chapter develops the formalism for both forces. Then it considers the point in time when the forces are balanced and the net force for universe expansion is zero. Then the universe may become static or it may proceed to expand further or it may contract. A static universe mirrors an elementary particle reflecting our view that universes are a form of particle.

The chapter ends by considering the possibility that the Hubble parameter and universe expansion are driven by the difference of Casimir force and the inner pressure due to mass-energy. *We treat the universe as a "bubble" in the Megaverse.*

We assume the universe may be viewed as a set of gambols as we earlier did successfully in a manner similar to the distribution of gambols within a fundamental particle. The universe *gambol temperature* T_g is related to the total mass-energy of the universe as the particle gambol temperature is related to the particle's mass:

$$kT_{gu} = c_g Ms = c_g M/(bt) \tag{7.1}$$

with

$$s = 1/bt$$

where $c_g = 0.0785$, k is Boltzmann's constant, s is the fractionation, $b = 8.67 \times 10^{-19}\ \text{sec}^{-1}$ by eq. 6.12, and M is the total mass-energy of the universe. We introduced a fractionation factor s that gives the gambol temperature T_{gu} a decreasing value with time.

At small time t the fractionation is large. At large time t the fractionation declines to a small value. This pattern will be used later to understand S_8 clumping.

7.1 Confinement Force on the Universe

We assume that gambols may be viewed as "universe-lets" that exist in the universe with its own vacuum. We further assume the Megaverse, within which the universe resides, has its own vacuum, which *differs* from the vacuum of the universe. The universe is treated as a three dimension sphere[61] both in itself and within the

[61] The time, and other Megaverse, dimensions are not part of the determination of vacuum energies at each instant.

Megaverse. The universe and Megaverse differ by the number of fundamental fermions: 256 in the universe and 1024 in the Megaverse. Based on their difference we find an inward directed (on the universe) Casimir force generated by the *difference between the Megaverse vacuum energy and the universe vacuum energy.*

7.2 Expansion Pressure of Universe Mass-Energy

The contents of the universe generate an expansion pressure (force) on the boundary of the universe.

$$P = nkT/V \tag{7.2}$$

using the Ideal Gas Law where n is the number of particles in the universe and P is the outward force per unit area. We assume the universe occupies a spherical volume $V_{universe} = 4/3\ \pi a^3$ with radius a.

$$P = 3nkT/(4\pi a^3) \tag{7.3}$$

The number of particles in the universe is estimated to be

$$\begin{aligned} n = \text{total mass of universe / proton mass} &= 1.712 \times 10^{81}/\text{GeV}/c^2 \,/\, 0.938\ \text{GeV}/c^2 \\ &= 1.82 \times 10^{81}\ \text{particles} \end{aligned} \tag{7.3a}$$

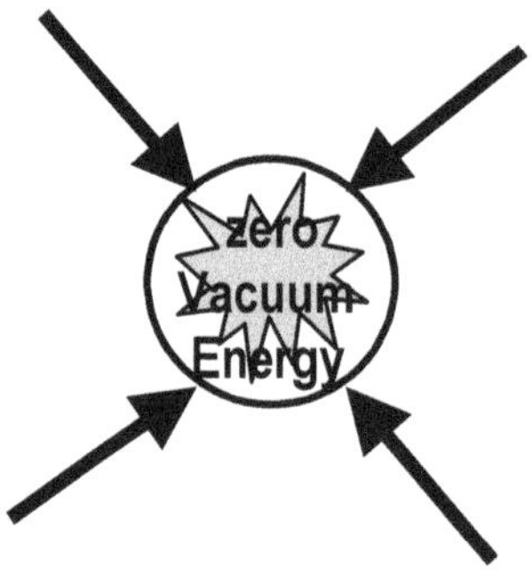

Figure 7.1. Schematic diagram of a universe's vacuum bubble showing the confining Casimir force from the exterior vacuum. This force is countered by internal gambol gas pressure P symbolized by a jagged interior star.

7.3 Gambol Model of the Expansion Pressure of Universe Mass-Energy

We choose to view the expansion pressure of the universe in terms of an ideal gas of gambols.[62]

---

[62] The universe confinement process proceeds analogously for a unioverse gambol model and a particle model. The difference between the universe and particle models appears when we consider the point in time when the universe pressure equals the Casimir force. Then universe expansion ends and it becomes "stable" like a particle.

$$P_g = 3nkT_{gu}/(4\pi a^3) \tag{7.4}$$

7.4 Energy within a Spherical "Bubble"

The vacuum of the universe has an energy due to its Dirac sea of fermions. We start with the total infinite Dirac fermion vacuum energy per unit volume summed for the 256 fundamental fermion species of our universe (Blaha number N = 7).

$$U_\infty = 256 \int_0^\infty d^3k\,(k^2 + M^2)^{½} \tag{7.5}$$

where M is the average mass.

We now consider a universe bubble of radius a within the Megaverse, with quantum fluctuations not present, in its Dirac fermion vacuum. The total vacuum energy per unit volume, using the approximation

$$d \approx 1/a \tag{7.6}$$

is

$$U_{\text{universe}} = 256 \int_d^\infty d^3k\,(k^2 + M^2)^{½} = 256 \int_{1/a}^\infty d^3k\,(k^2 + M^2)^{½} \tag{7.7}$$

or

$$U_{\text{universe}} = \infty - 256\{4\pi[d(d^2 + M^2)^{3/2}/4 - dM^2(d^2 + M^2)^{½}/8 - \text{arsh}(d/(96M^3)]\} \tag{7.8}$$

per unit volume. The infinite term is not relevant as will be seen when we take the derivative of U_{universe}. For $d >> M$ we find (neglecting the infinity)

$$U_{\text{universe}} \approx -256\pi d^4 \tag{7.9}$$

$$\approx -256\pi a^{-4} \tag{7.10}$$

per unit volume for our universe.

Multiplying by the volume of the bubble we find a total finite vacuum energy for our universe is

$$U_{\text{universe}} \approx -1024\pi^2 a^{-1}/3 = -1024\pi^2 d/3 \tag{7.11}$$

7.5 Vacuum Energy in the Six Space-Time Dimension Megaverse

We find the total bubble energy for the six space-time Megaverse subspace (Blaha Number N = 6) of our universe is

$$U_{\text{Mega}} \approx -4 \times 1024\pi^2 a^{-1}/3 = -4096\pi^2 d/3 \tag{7.12}$$

using 1024 as the number of fundamental fermions in the Megaverse.

Subtracting the energy of our universe from the Megaverse energy we find

$$U_{net} = U_{totMega} - U_{universe} = -3072\ \pi^2 d/3 = -1024\ \pi^2 d \qquad (7.13)$$
$$= -1024\ \pi^2/a$$

7.6 Casimir Effect Confining Gambol Force

This section calculates the force on a particle due to the Casimir effect.[63] The inward pressure[64] (force) on the universe bubble boundary of area, $4\pi a^2$, is

$$P_{Casimir} = -(4\pi a^2)^{-1}\ \partial U_{net}/\partial a \qquad (7.14)$$
$$= 256\pi\ a^{-4}$$

It is directed inward as shown in Fig. 7.1. It is counterbalanced by gambol thermodynamic pressure. Since the radius a is presumably ultra-small the Casimir pressure will be extremely large. Thus gambols are confined except, perhaps, for "ultimate" energies exceeding the Planck energy.

7.7 Net Force Gambol Confinement – Universe Pressure

We now find the difference of the outward pressure of the gambol gas and the inward Casimir vacuum force:

$$F_{net} = 256\pi\ a^{-4} - nkT_{gu}/(btV) \qquad (7.15)$$

where T_{gu} is defined by eqs. 7.1 and 7.4, and where M is the total mass-energy of the universe. Substituting the volume

$$F_{net} = 256\pi\ a^{-4} - \tfrac{3}{4}\ nc_g M/(\pi bt\ a^3) \qquad (7.16)$$

As in the particle confinement calculation of Blaha (2023e) we set n by eq. 7.3a, the fractionation. Consequently

$$F_{net} = 256\pi\ a^{-4} - 5.8 \times 10^{97}\ M/(ta^3) \qquad (7.17)$$

7.8 When Will the Expansion Force be Zero?

When the Casimir force equals the mass-energy pressure the time is $t = t_{Zero}$ satifies

$$256\pi\ a^{-4} = 5.8 \times 10^{97}\ M/(t_{Zero} a^3) \qquad (7.18)$$

or

$$t_{Zero} \approx 5.8 \times 10^{97}\ aM / (256\ \pi) = 7.2 \times 10^{94}\ aM \qquad (7.19)$$

Note: aM << 1. The value of t_{Zero} is far more than the current universe time $t_{NOW} = 4.35 \times 10^{17}$ sec.

---

[63] H. G. B. Casimir, Koninkl. Ned. Adak. Wetenschap. Proc. **51**, 793 (1948), S. K. Lamoreaux, Phys. Rev. Lett. **78**, 5 (1997), L. S. Brown and G. J. McClay, Phys. Rev. **184**, 1272 (1969), and references therein.

[64] This result is comparable to the often quoted Casimir example of the force between two parallel conducting plates due to the quantum fluctuations of the electromagnetic field between the plates.

7.9 A Pressure Driven Hubble Parameter?

The driving force for universe expansion represented by eq. 7.17 raises the question whether it might be the driving force for Hubble expansion if we treat a as the scale factor. We consider this possibility now with

$$dH/dt = \text{Force} = F_{net}(a) \quad (7.20)$$

noting H has the dimension sec^{-1} and dH/dt has the dimension sec^{-2} of acceleration. We separate $F_{net}(a)$ into two parts by treating a = a(t) as the scale factor satisfying

$$H(t) = d/dt \ln a(t) \quad (7.21)$$

7.9.1 For small a

In this case where t is less than the Interregnum minimum 10^{14}: in Fig. 6.2 we see:

$$F_{net-}(a) = 256\pi\, a^{-4} = d^2/dt^2 \ln a(t) \quad (7.22)$$

Letting

$$a(t) \sim \alpha t^{\beta} \quad (7.23)$$

We see

$$256\pi\alpha^{-4}t^{-4\beta} = \beta\, d^2/dt^2 \ln t = -\beta/t^2$$

with the results

$$\beta = 1/2 \quad (7.24)$$
$$\alpha = -(512\pi)^{1/4} = -6.33$$

Thus the Casimir force, by itself, yields a negative contribution to the scale factor. The β power corresponds to the scale factor power in the radiation phase[65] of universe expansion where

$$a(t) \sim (2\Omega_\gamma^{1/2}H_0)^{1/2}t^{1/2} \quad (7.25)$$

7.9.2 For large a

In this case where t is greater than the Interregnum maximum 10^{17} we see:

$$F_{net+}(a) = -5.8 \times 10^{97}\, M/(ta^3) = d^2/dt^2 \ln a(t) \quad (7.26)$$

Letting

$$a(t) \sim \alpha t^{\beta}$$

then

$$5.8 \times 10^{97}\, M/(\alpha^3 t^{3\beta+1}) = \beta/t^2$$

with the results

$$\beta = 1/3 \quad (7.27)$$
$$\alpha = 9.93 \times 10^{60}\ \text{GeV}/c^2$$

The β power[66] 1/3 corresponds to ½ of the scale factor power in the matter-dominated phase[67] of universe expansion where

---

[65] See p. 208 of Blaha (2004).

$$a(t) \sim (2\Omega_m^{1/2} H_0)^{2/3}\, t^{2/3} \tag{7.28}$$

The mass-energy pressure, in itself, yields a positive contribution to the scale factor.
If the gambol temperature (eq. 7.1) is changed to[68]

$$kT_{gu} = c_g M\,(1 + bt)/(bt) \tag{7.29}$$

giving the universe a finite non-zero temperature, $kT_{gu} = c_g M$, as $t \rightarrow \infty$ then

$$\beta \rightarrow 2/3 \tag{7.30}$$

yielding the form of eq. 7.28, which was obtained from a consideration of the a(t) scale factor Einstein equation. The $t \rightarrow \infty$ constant value of T_{gu} may be a result of quantum effects – universe *zitterbewegung.*

It appears there is a reality to the driving force for universe expansion based on a confining universe Casimir force and the pressure due to the mass-energy of the universe. The success of the approach gives strong support for:

1. *The particle interpretation of universes: creation as point objects in a higher space, expansion through a "mismatch" with internal mass-energy pressure larger than Casimir confinement, and a strong similarity to gambol confinement in particles.*

2. *The existence of a Megaverse within which our universe resides.*

3. *Our universe confined by a Casimir vacuum force.*

4. *A wider theory of the universe's evolution beyond the current Standard Models of Cosmology.*

5. *A gambol model of universes similar to the gambol model of particles: quarks, leptons and hadrons. This is further evidenced by the Gambol S_8 Model in chapter 8.*

The success of the gambol Hubble parameter calculations here, and the success in chapter 6 and in the S_8 gambol model support the view of our universe as a type of particle in the Megaverse within the framework of Cosmos Theory spaces. Our universe has a Blaha number N = 7 space and the Megaverse has a Blaha number N = 6 space.

---

[66] The explicit factor of t in eq. 7.15 changed the power from 2/3 to 1/3 above.
[67] See p. 205 of Blaha (2004).
[68] The (1 + bt)/bt factor is well approximated by 1/t for $t < t_{NOW}$ since $bt << 1$ for times before $bt_{NOW} = 0.29$.

8. Universe S_8 Clumping

There is experimental evidence suggesting the mass-energy of the universe experiences a form of clumping that evolves over time. The Clumping Index S_8 has a current value estimated to be between $S_8 = 0.74$ and 0.83 by experimental groups. We will use the value:

$$S_{8NOW} = 0.775 \tag{8.1}$$

We develop a Gambol S_8 Model by assuming the clumping at the Big Bang point is zero, and that it increases over time "linearly" to the current value. The gambol fractionation approach appears to meet our goal: We set

$$S_8 = C/s = Cbt \tag{8.2}$$

where C is a constant, b is the parameter defined in chapter 6, and t is the time.
Our model is based on the concept:

> Consider a liquid of two species thoroughly mixed initially. As time progresses the liquid separates into clumps of one species surrounded by the liquid of the other species. Thus it exhibits an S_8 type of clumping. S_8 may be viewed as measuring the amount of "coarse graining" or fractionation.

The combination of Gambol Quantum Field Theory, the Gambol Hubble Model, the gambol and vacuum energy based calculation of universe confinement, and the Gambol S_8 Model developed here strongly suggests the particle interpretation of universes is correct and it supports Cosmos Theory.

8.1 First Gambol Clumping Model

We assume S_8 clumping reflects universe gambol fractionation with large fractionation s for small time t. and small fractionation for large time. Thus we define a linear clumping index S_8 with

$$S_8 = C/s = Cbt \tag{8.1}$$

We determine C using the S_8 value 0.775. Since $bt_{NOW} = 0.29$ we find

$$C = 2.67 \tag{8.2}$$

8.2 Clumping Parameter S_8 at t = 1 sec

For t = 1 sec we find

$$S_8 = 6.67 \times 10^{-19} \tag{8.3}$$

and the fractionation is the very large

$$s = 4.33 \times 10^{17}$$

8.3 The Time when s = 1

The time t_l when the fractionation s = 1 is determined by

$$s = 1 = 1/bt_l$$

$$t_l = 1.5 \times 10^{18} \text{ sec} = 3.45\ t_{NOW} \tag{8.4}$$

The universe becomes one "clump" of a more or less uniform density at $t = t_l$.

8.4 Relation of S_8 to Gambol Temperature kT_{gu}

The universe gambol temperature is

$$kT_{gu} = 0.0785\ M/bt \tag{8.5}$$

where $M = 1.712 \times 10^{81}\ \text{GeV/c}^2$ is the total mass-energy of the universe. Comparing to S_8 we see the product of kT_{gu} with S_8 is a constant

$$\begin{aligned} S_8\, kT_{gu} &= 0.0785\ CM \\ &= 0.209\ M = 3.6 \times 10^{80}\ \text{GeV/c}^2 \end{aligned} \tag{8.6}$$

8.5 Second S_8 Model Related to Gambol Mass-Energy Model of Chapter 4

Assuming the clumping is roughly proportional to the mass-energy of the clump, we find $S_8(t)$ is proportional to the ratio of mass-energy distribution value to the total mass-energy. We can use the Mass-Energy Gambol Model of chapter 4 to create another S_8 Clumping Index Model.

$$S_8(t) = N_{S8}\, E_\varepsilon(t)/M_u \tag{8.7}$$

where $E_\varepsilon(t)$ is derived in chapter 4 and $N_{S8} = 3.5942$. See Figs. 8.1 and 8.2.

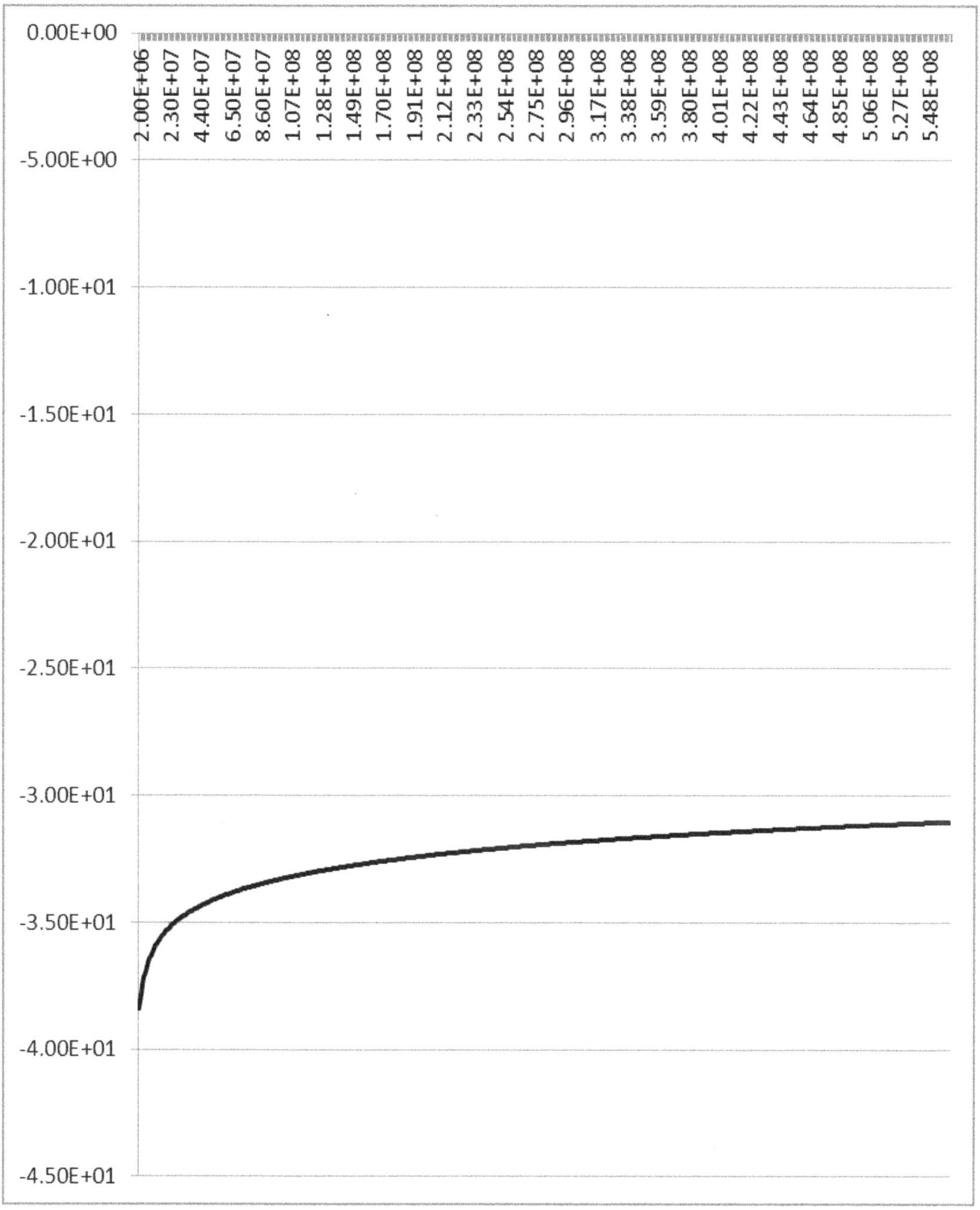

Figure 8.1. Plot of Logarithm base 10 of S_8 in eq. 8.7 for early times in seconds since the Big Bang.

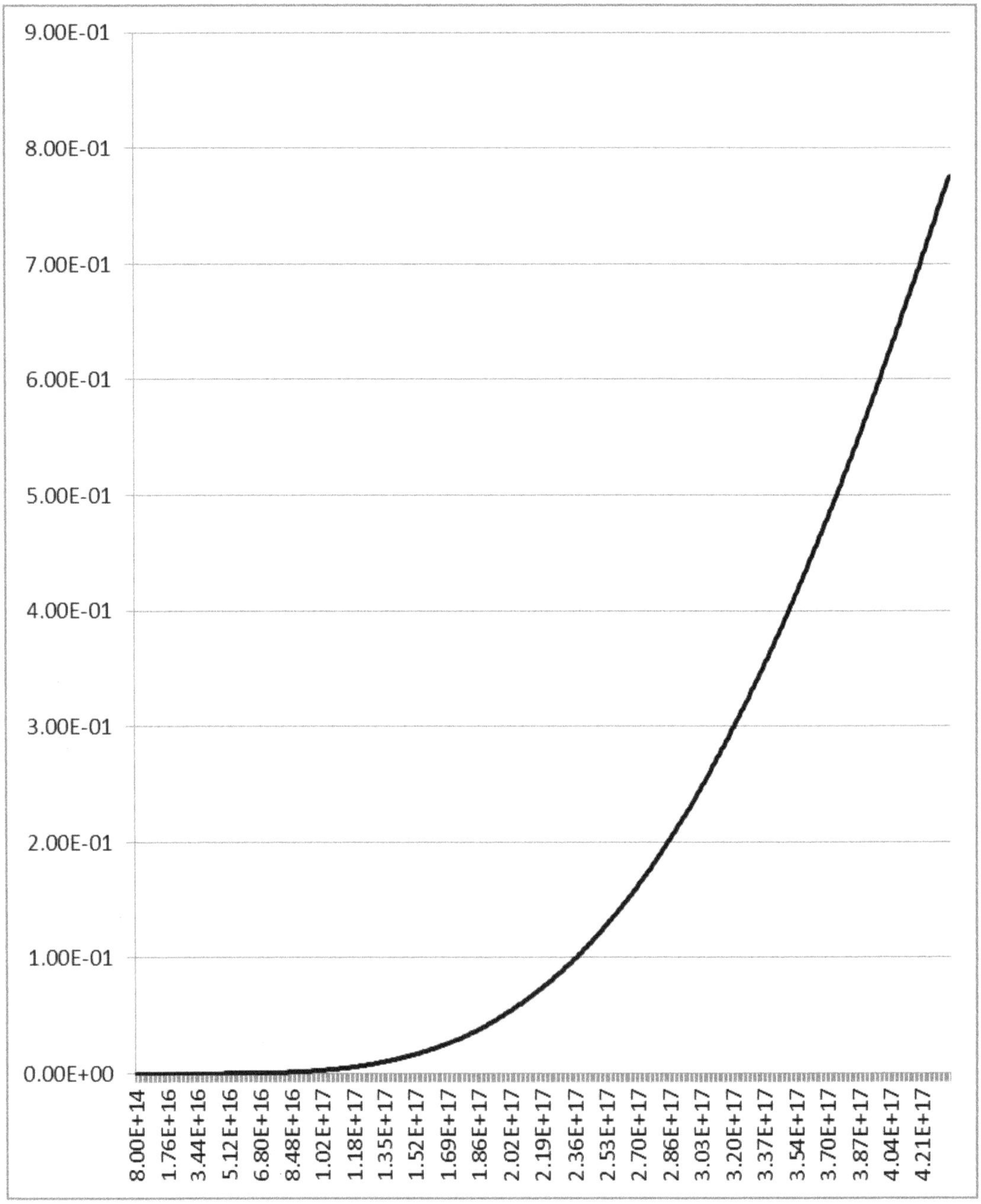

Figure 8.2. Plot of Logarithm base 10 of S_8 in eq. 8.7 for recent times in seconds since the Big Bang.

8.6 Comments

We conclude the S_8 clumping parameter appears to be related to the fractionation and to the universe gambol temperature. Clumping reflects gambol size. Smaller clumping near $t = 0$ yields a uniform universe density which is consistent with the view that the universe is very small and very hot then. There is larger clumping near $t = t_{NOW}$ reflecting the much less uniform density and temperature.

The universe gambol size (which is a measure[69] of clumping) with $1/s = bt$ increases with time t as the universe expands (chapter 4). The universe becomes more coarse grained with larger clumps. It becomes ultimately one clump of uniform density in the large. The Virgo Supercluster reflects the large clustering with its 100,000 galaxies.

The universe begins small and is of a uniform distribution made of infinitesimal clumps. It then progresses to become larger and to have larger clumps (stars and proto-galaxies, galaxies, clusters, and so on). There are large clumps of galaxies connected by cosmic vines. Ultimately the universe becomes one tremendous clump.

[69] Bigger gambol size corresponds to larger clumps. A clump is more or less a gambol of roughly constant, larger density.

9. The Reality of Gambols

Gambols are abstract entities whose existence is justified by their consequences. They are not real particles.

We view gambols as a theoretical construct. A particle, universe, or other entity may be treated in different ways using different gambol models with each model suited to a specific purpose.

Each gambol fractionation is theoretical. So one cannot say that there is one and only one specific fractionation into a specific set of "real" gambols. The gambol concept is thus more abstract than the quark concept since quarks, although confined, are real – meaning we can directly measure them as particles> A hadron is always composed of a specific number of quarks.[70]

Gambols adopt the characteristics of their parent particle. Gambols appear in sets of identical gambols.[71] Creation operators project out an individual gambol to represent a particle in an interaction. The possibility of constructs like the gambol construct, where there is a precursor entity that is not directly experimentally measureable but leads to experimentally measurable results, was considered by 20th Century Philosophers of Science such as R. B. Braithwaite (*Scientific Explanation*, Cambridge University Press, 1953) and Karl Popper whose thesis was that Physical theories can never be proven but can be falsified and shown to be true or false via experiment.

In this chapter we outline some of the possible types of gambol models.

9.1 Gambols in Particles – Mass and Momentum

In Blaha (2023e) we described particle decay and hadron interactions in a gambol model that was based on gambols having a fractionated mass and momentum. Chapter 3 described the quantum field theory of the gambol model of particles, hadrons and hadron interactions where gambols have the internal symmetry and spin of the parent particle.

9.2 Gambols in Universes and Megaverses

In chapters 4 – 11 we describe gambol models of universes and Megaverses. Chapters 4, 5, 7 and 8 use Planckian distributions for gambols. Chapter 6 uses an anti-Planckian distribution to determine the Hubble Parameter time dependence.

Thus entities may have different gambol models to suit different purposes. The net result is theoretical gambols – not Physical in the sense of leptons and quarks.

---

[70] Modulo a possible quark sea.

[71] They evade the Pauli Exclusion Principle as we described earlier in chapters 3 and 4.

9.3 Characterized by Internal Symmetry fractionation

It is possible to fractionate the internal symmetry groups to create gambols. We present below an extract from Blaha (2022d):

12.9 CASe Groups: SU(n), su(n, n), U(1/n)

The groups associated with the HyperCosmos all have group numbers n or 1/n that are powers of 2. Thus they all support fractionation.

12.9.1 Fractionating CASe Groups: su(n, n)

The CASe groups that appear in Fig. 12.1 of the form of su(n, n) have 4n real-valued coordinates in their fundamental irreducible representations. It is important to note that n is necessarily a power of 2 due to its relation to Cayley numbers. Because n is always even, and a power of 2, in HyperCosmos CASe groups, we can fractionate su(n,n) to su(n/2,n/2) and thence down to su(1,1) Then we can further fractionate it into the fractional groups U(1/n) in Fig. 12.1 that we discuss next.

The basis of the fractionation is the number of coordinates in the fundamental irreducible representation of a CASe group, which is a power of 2. Fractionation yields a CASe group with *half the number of coordinates* at each stage of fractionation. Thus the dimensions of the group representation is fractionated down to the su(1, 1) fraction.

The su(1, 1) representation has four real-valued coordinates. The reduction of su(1,1) therefore produces the Blaha number 10 U(1) CASe group as Fig. 12.1 illustrates. In the next section we describe fractionation for CASe U(1/n) groups.

12.9.2 Fractionating CASe Groups: U(1/n)

The U(1) group elements may be represented as:

$$e^{i\theta}$$

where θ ranges from 0 to 2π. Reducing the U(1) representation to U(½) requires a reduction in the range of coordinates. This process is accomplished by modifying the representation to

$$e^{i\theta/2}$$

The θ range remains 0 through 2π. Thus the resulting operators are factored by 2. A similar procedure leads to the representation of U(1/n) where n is a power of 2:

$$e^{i\theta/n}$$

Thus the CASe groups U(1/n) support fractionation of the coordinates of their group irreducible representations.[72]

12.9.3 Fractionating Symmetry Groups: SU(n)

HyperCosmos spaces have a dimension array that can be viewed as the fundamental representation of SU(n) where $n = d_{dn}/2 = 2^{21-2N}$ for some N. The dimension array group is separated into sets of symmetry groups in each HyperCosmos space. The symmetry groups are copies of color SU(4) assuming lepton-quark doublets before symmetry breaking and into SU(2)⊗U(1) ElectroWeak groups.

If we consider SU(n) in general we note that it can be fractionated indefinitely if n is a power of 2. U(1) also can be fractionated indefinitely. Thus the groups of interest in the set of

---

[72] One could extend this discussion to U(1/a) where a is a real number.

symmetry groups of each space (universe) are exactly the ones susceptible to fractionation based on powers of 2^n.

SU(n), for other values of n, does not support an indefinite powers of 2 fractionation with the sole exception being U(1). For odd n the fundamental representation coordinates after the first fractionation become fractions n/m that are not equal to powers of 2. Thus it would not be consistent with the 2^n fractionation of the HyperCosmos and thus not support a fractional dynamics based on 2^n fractionation.

Thus one can use a product of fractionated groups to represent a group with a subgroup product. For example,

$$\text{SU(1) has the fractionated subgroup product } 1/\otimes \prod_{\substack{n=2 \\ n \text{ even}}}^{\infty} \otimes U(1/n).$$

if we treat ⊗ as a symbolic operator. An SU(N) group has a subgroup $[SU(1)^N$ supporting the iteration of the SU(1) factorization above. SU(N) supports factorization.

9.4 Characterized by Spin Fractionation

Similarly one may fractionate a spin group such as SL(2,**C**).

9.5 Characterized by Other Features – Quark Core Example

Other entity features may be used to define a gambol structure. For example chapter 12 applies a gambol structure to the quark cores of neutron stars based on radial distance fractionating from the center of the quark core.

Chapter 11 uses the Cayley-Dickson number to define a gambol structure for the ten universes of the Cosmos.

9.6 Combinations for Gambol Fractionation

Combinations of the above fractionations may be created for special purposes in the study of particle and universe features and interactions.

10. Cosmos Sequence of Universes as Gambols

Cosmos Theory has a sequence of ten HyperCosmos spaces that may be used to generate a sequence of universes.[73] We may view the universes as each being generated in turn from an 18 space-time dimension universe, called the Parent universe, with Blaha Space Number 0.

In this chapter we will consider generated universes to be gambols. The Parent universe has a set of gambols, one of which becomes a 16 space-time dimension universe. This gambol universe has a set of gambols, one of which becomes a 14 space-time dimension universe. And so on.

The Parent universe is thus fractionated into gambol universes. The fractionation into a 1/s fraction of the Parent universe gives a gambol universe with

$$s = 2^{2N}$$

for Blaha Number N. The space-time dimension r is

$$r = 18 - 2N$$

The lowest fermion spin is

$$\text{spin} = r/4 - ½$$

In our universe the lowest fermion spin is ½.

The dimension array d_{dN} has

$$d_{dN} = 2^{22-2N} \text{ dimensions}$$

We view it as a square array of 2^{11-N} by 2^{11-N} dimensions. For our universe which is a HyperCosmos N = 7 space, there are four space-time dimensions, $s = 2^{14}$, and there is a dimension array with $d_{d7} = 2^8 = 256$ dimensions (as in the author's Unified SuperStandard Theory.) See Fig. 10.1 for the list of spaces and dimensions.

The chain of gambol universes has an 18 space-time dimension "Parent" quantum field with coordinates y_0 containing nine nested gambol universe quantum fields with a form:

$$\Psi(y_0, y_1, y_2, y_3, y_4, y_5, y_6, y_7, y_8, y_9) \quad (10.1)$$

We may separate the sequences of gambols by using sequences of parent – gambol fields of the form

$$\Psi_i(y_i, y_{1i+1}) \quad (10.2)$$

[73] Chapter 11 presents a Planckian distribution for a larger set of universes.

for i = 0, 1, …, 8. The form of each of their Lagrangians and field theories has the same form as the Megaverse – universe gambol model of chapter 5. They may be made multi-universe at any stage using the approach of section 5.8.

10.1 The Complete Parent Universe and Sequence of Gambol Universes

It is possible to formulate the theory using a generalization of the creation and annihilation operators of chapter 5. The Parent universe quantum fields generate gambol universe quantum fields. The states of the gambol universe fields can be used to define universes as we will see later. Then a set of universes may be created for each type of HyperCosmos space. Chapter 11 describes the set of universes that might exist that are the result of probabilistic transformations governed by a Planckian distribution.

As previously, we view universes as a form of particle.

We begin by defining a free Parent PseudoFermion PseudoQuantum Lagrangian with two quantum fields ψ_1 and ψ_2 with the form of eq. 10.1 that are functions of ten sets of coordinates $y_0, y_1, \ldots y_{09}$. The free Lagrangian is[74]

$$\mathscr{L} = \overline{\psi}_{2\alpha_0\alpha_2 \ldots \alpha_9}[-M^{-1}\prod_{i=0}^{9} \gamma_{i\alpha_i\kappa_i}{}^{\mu_i}\partial/\partial y^{\mu_i} - M]\psi_{1\kappa_0\,\kappa_2 \ldots \kappa_9 M} +$$

$$+ \overline{\psi}_{1\alpha_0\alpha_2 \ldots \alpha_9}[-M^{-1}\prod_{i=0}^{9} \gamma_{i\alpha_i\kappa_i}{}^{\mu_i}\partial/\partial y^{\mu_i} - M]\psi_{2\,\kappa_0\,\kappa_2 \ldots \kappa_9 M} \tag{10.2}$$

where $\gamma_i{}^{\mu}$ is a Dirac matrix for the coordinates y_i in the r_i dimension space-time of Blaha Space Number i space's universe, M is the Space Number 9 space's mass, and

$$\overline{\psi}_{i\,\alpha_0\alpha_2 \ldots \alpha_9 M} = \psi_{i\,\kappa_0\,\kappa_2 \ldots \kappa_9 M}{}^{\dagger}\prod_{i=0}^{9} \gamma_{\kappa_i\alpha_i}{}^{0} \tag{10.3}$$

for i = 1, 2.

Following a development similar to eqs. 5.2 – 5.14 we come to the quantum field:

$$\psi_{i\,\alpha_0\alpha_2 \ldots \alpha_9\gamma M}(y_0,y_1,\ldots,y_9)=\sum_{s}\sum_{spins} \{ \prod_{k=0}^{9} [\int dp_k{}^{r_k-1}]\, \mathscr{N}(p_0,p_1, \ldots ,p_9)\cdot$$

$$\cdot[b_{\gamma iM}(s, p_0, p_1, \ldots p_9, s_0,s_1,s_2, \ldots s_9) \prod_{k'=0}^{9} u_{\alpha_{k'}}(p_{k'},s_{k'}) \exp(-i \sum_{k''=0}^{9} p_{k''}\cdot y_{k''}) +$$

$$+ d_{\gamma iM}(s, p_0, p_1, \ldots p_9, s_0,s_1,s_2, \ldots s_9) \prod_{k'=0}^{9} v_{\alpha_{k'}}(p_{k'},s_{k'}) \exp(i \sum_{k''=0}^{9} p_{k''}\cdot y_{k''})]\} \tag{10.4}$$

plus Hermitean conjugates for i = 1, 2 where $\mathscr{N}(p_0,p_1, \ldots ,p_9)$ is a normalization factor.

---

[74] This derivation has a parallel derivation for boson universes. See chapter 14.

The composite creation and annihilation operators satisfy the anti-commutation relations:

$$\{b_{\gamma iM}(s, P, S), b_{\gamma' jM}(s', P', S')^{\dagger}\} = \{d_{\gamma iM}(s, p, S), d_{\gamma' jM}(s', p', S')^{\dagger}\} =$$
$$= (1 - \delta_{ij})\,\delta_{\gamma\gamma'}\delta_{ss'}\,\delta_{SS'}\,\delta^{R-1}(\mathbf{P} - \mathbf{P'})U_g(\varepsilon(s,q)) \qquad (10.5)$$

where P represents the momentum arguments $p_0, p_1, \ldots p_9$, S represents the spin arguments $s_0, s_1, s_2, \ldots s_9$ and U_g is given by eqs. 2. 22. The factors $\delta_{SS'}\delta^{R-1}(\mathbf{P} - \mathbf{P'})$ represent

$$\prod_{k=0}^{9} \delta_{S_k S_k'}\, \delta^{r_k - 1}(\mathbf{p_k} - \mathbf{p_k'}) \qquad (10.6)$$

The other composite anti-commutation operators yield zeros:

$$\{b_{\gamma iM}(s, P, S), b_{\gamma' jM}(s', P', S')\} = 0 \qquad (10.7)$$
$$\{b_{\gamma iM}(s, P, S)^{\dagger}, b_{\gamma' jM}(s', P', S')^{\dagger}\} = 0$$
$$\{d_{\gamma iM}(s, P, S), d_{\gamma' jM}(s', P', S')\} = 0$$
$$\{d_{\gamma iM}(s, P, S)^{\dagger}, d_{\gamma' jM}(s', P', S')^{\dagger}\} = 0$$

and anti-commutators of b and d type operators are zero as well.

10.2 Gambol Universe Operators

We now determine the gambol universe creation/annihilation operators for the gambol universes labeled with Blaha Number N > 0. (See Fig. 10.1.) The N = 0 universe is the Parent of all gambol universes.

We define the gambol universe operator for Blaha Number N = k with

$$b^{1/s}{}_{g\gamma iM}(s, p_k, s_k) = \sum_{\substack{s_m \\ m=0 \\ m \neq k}}^{9} \Big[\prod_{\substack{j=0 \\ j\neq k}}^{9} \big(\int d^{r_j - 1}p_j \mathcal{N}'(P)\big)^{1/2}\Big]\, b_{\gamma iM}(s, P, S) \qquad (10.8)$$

and similarly its complex conjugate where there is a sum over all spins except the kth spin, where $s = 2^{2N} = 2^{2k}$ is the fractionation, $\mathcal{N}(P)$ is the normalization and i is the PseudoQuantum index with i = 1, 2.

The gambol universe field operator anti-commutators are:[75]

$$\{b^{1/s}{}_{g\gamma iM}(s, p_k, s_k), b^{1/s'}{}_{g\gamma' jM}(s', p_k', s_k')^{\dagger}\} = (1 - \delta_{ij})\,\delta_{s,s'}\delta_{s_K s_K'}\delta_{\gamma,\gamma'}\delta^{r_K - 1}(\mathbf{p_k} - \mathbf{p_k'})\, U_g(\varepsilon(s,p_k))$$

and

$$\{d^{1/s}{}_{g\gamma iM}(s, p_k, s_k), d^{1/s'}{}_{g\gamma' jM}(s', p_k', s_k')^{\dagger}\} = (1 - \delta_{ij})\,\delta_{s,s'}\delta_{s_K s_K'}\delta_{\gamma,\gamma'}\delta^{r_K - 1}(\mathbf{p_k} - \mathbf{p_k'})\, U_g(\varepsilon(s,p_k)) \qquad (10.9)$$

where eqs. 2.22 specifies $U_g(\varepsilon(s,q))$ with m = M.

---

[75] This development parallels that of chapter 5.

Eqs. 3.18 and 3.19 have fractional integrations that utilize

$$[(\int dp^{r-1}\,\mathscr{N}(p,q))^{1/2}]^2 = \int dp^{r-1}\,\mathscr{N}(p,q) \tag{10.10}$$

as in Riemann-Liouville integrals. Note γ and γ' are internal symmetry indices.

The other anti-commutators are zero:

$$\begin{aligned}
&\{b^{1/s}{}_{g\gamma iM}(s, p_k, s_k)^\dagger, b^{1/s'}{}_{g\gamma' jM}(s', p_k', s_k')^\dagger\} = 0 \\
&\{d^{1/s}{}_{g\gamma iM}(s, p_k, s_k)^\dagger, d^{1/s'}{}_{g\gamma' jM}(s', p_k', s_k')^\dagger\} = 0 \\
&\{b^{1/s}{}_{g\gamma iM}(s, p_k, s_k), b^{1/s'}{}_{g\gamma' jM}(s', p_k', s_k')\} = 0 \\
&\{d^{1/s}{}_{g\gamma iM}(s, p_k, s_k), d^{1/s'}{}_{g\gamma' jM}(s', p_k', s_k')\} = 0
\end{aligned} \tag{10.11}$$

and anti-commutators for combinations of b and d type operators are zero as well.

The gambol creation/annihilation operators above can be used to define the kth gambol universe fermion quantum fields:

$$\begin{aligned}
\psi^{1/s}{}_{g\gamma M1}(x) &= \Sigma_{p_k}\,[b^{1/s}{}_{g\gamma 1M}(p_k, s_k)\, f_p(x) + d^{1/s}{}_{g\gamma 1M}(p_k, s_k)^\dagger f_p^*(x)] \\
\psi^{1/s}{}_{g\gamma M2}(x) &= \Sigma_{p_k}\,[b^{1/s}{}_{g\gamma 2M}(p_k, s_k) f_p(x) + d^{1/s}{}_{g\gamma 2M}(p_k, s_k)^\dagger\, f_p^*(x)]
\end{aligned} \tag{10.12}$$

where $k \neq 0$ and the indices 1 and 2 are PseudoQuantum indices.

10.3 Parent Universe Creation/Annihilation Operators

The Parent universe operators may be defined using the composite creation/annihilation operators above. We replace a sum[76] over s with a sum over energies ε so as to take advantage of the normalization sum in eq. 2.21:[77]

$$b_{\gamma iM}(p_0, s_0) = \int_0^\infty d\varepsilon\, \sum_{m=1}^{9}{}_{\text{spins}}\ [\prod_{j=0}^{9} (\int d^{r_j - 1}p_j\, \mathscr{N}(P))^{1/2}]\, b_{\gamma iM}(s(\varepsilon), P, S) \tag{10.13}$$

plus complex conjugate, where s_0 is the Parent universe quantum field spin.

The anti-commutation relations are[78]

$$\{b_{\gamma iM}(p_0, s_0), b_{\gamma' jM}(p_0', s_0')^\dagger\} = (1 - \delta_{ij})\, \delta_{s_0,s_0'}\delta_{\gamma\gamma'}\delta^{r_0-1}(\mathbf{p_0} - \mathbf{p_0'}) \tag{10.14}$$

using eq. 3.21 with $\varepsilon(s)$ and $\varepsilon(s')$. We make the s (and ε) sums into integrations for analytic convenience just as the discrete sum over hν in the black body Planck distribution derivation is similarly made continuous.

The other anti-commutation relations are:

$$\{d_{\gamma iM}(p_0, s_0), d_{\gamma' jM}(p_0', s_0')^\dagger\} = (1 - \delta_{ij})\, \delta_{s_0,s_0'}\delta_{\gamma\gamma'}\, \delta^{r_0-1}(\mathbf{p_0} - \mathbf{p_0'}) \tag{10.15}$$

---

[76] The sum over s begins as a discrete sum over powers of 2 according to Limos. We replace it with a continuous value for s, which makes the transition to an integral over ε possible.

[77] Using a sum over ε supports eq. 5.13.

[78] The s, s' and $\delta_{ss'}$ terms are equivalents of ε energie terms due to the 1:1 relation of s and ε specified in eq. 2.23.

$$\{b_{\gamma iM}(p_0, s_0), b_{\gamma' jM}(p_0', s_0')\} = 0$$
$$\{d_{\gamma iM}(p_0, s_0), d_{\gamma' jM}(p_0', s_0')\} = 0$$
$$\{b_{\gamma iM}(p_0, s_0)^\dagger, b_{\gamma' jM}(p_0', s_0')^\dagger\} = 0$$
$$\{d_{\gamma iM}(p_0, s_0)^\dagger, d_{\gamma' jM}(p_0', s_0')^\dagger\} = 0$$

and the anti-commutators of b and d type operators are zero as well.

The creation/annihilation operators above are those that appear in the Parent universe quantum field. Thus we may define a Parent universe quantum field with

$$\psi_{\gamma M1}(x) = \Sigma_p\, [b_{\gamma 1M}(p, s_0)\, f_p(x) + d_{\gamma 1M}(p, s_0)^\dagger f_p{}^*(x)] \quad (10.16)$$
$$\psi_{\gamma M2}(x) = \Sigma_p\, [b_{\gamma 2M}(p, s_0) f_p(x) + d_{\gamma 2M}(p, s_0)^\dagger\, f_p{}^*(x)]$$

with internal symmetry index γ for the Parent coordinates, spin s_0 and momentum.

10.4 Form of the Descent of Gambol Universes

The set of nine gambol universes generated from the Parent universe quantum field have a specific interrelated set of properties as outlined in Fig. 10.1. The essential feature of the set of gambol universes is the quartering of symmetry dimension arrays d_{dN} level by level as N increases.

10.4.1 Space-Time Dimensions

The space-time dimensions *decrease* by two as N increases by one. This feature is the result of the impact of the creation/annihilation operator counting to form the basis of the Cosmos Spaces (seen in earlier books by the author.) Odd dimensions are excluded because these operators are the same for odd dimensions and the even dimensions below them. (Including odd number dimensions would constitute a form of double counting.) The number of space-time dimensions is restricted to the range 0 through 18. Negative space-time dimensions are excluded. An arbitrary cutoff at 18 was made. Chapter 11 provides a probabilistic mechanism for the full set of 23 Cosmos Theory universes that limits the maximum space-time dimension to 18.

One may view the coordinate space of gambol N as a subspace of the coordinate space of gambol N – 1.

10.4.2 Fermion Spins

The lowest fermion spin in each gambol universe is determined by its space-time dimension.

10.4.3 Internal Symmetries

The size of the dimension array of a Blaha Number N space is one-quarter the dimension array of the Blaha Number N – 1 space:

$$d_{dN} = \tfrac{1}{4}\, d_{dN-1} \quad (10.17)$$

and

$$d_{dN} = 2^{22-2N}$$

This feature of dimension arrays was presented in detail in earlier books by the author. Fig. 10.1 displays it.

Since one may view dimension arrays as square arrays, the N dimension array may be viewed as one quadrant of the N – 1 dimension array.

The range of γ index values is given by eq. 10.17, decreasing by a factor of 4 as one goes from level N to N + 1.

10.5 Gambol Universe States

Each gambol quantum field may define one or more universes. A single universe of the type of space s (or N) is defined by

$$|1> = b^{1/s}{}_{g\gamma 2M}(s, p, s_0)^\dagger|0> \tag{10.18}$$

where s_0 is the spin and p is its momentum in the Parent universe.

Multiple universes may be defined in the Parent universe. We express a k universe quantum state within the Parent universe using k universe gambol creation operators:

$$|g\ k> = b^{1/s_1}{}_{g\gamma 2M}(s_1, p_1, S_1)^\dagger\ b^{1/s_2}{}_{g\gamma 2M}(s_2, p_2, S_2)^\dagger|0> \ldots\ b^{1/s_k}{}_{g\gamma 2M}(s_k, p_k, S_k)^\dagger|0> \tag{10.19}$$

\

where S_i is the universe i spin for i = 1, 2, … k. The fractionations s_i may or may not be the same. The fractionations, momentum spin combinations (s_i, p_i, S_i) must be distinguishable due to eq. 2.9.

Our universe is a gambol universe generated probabilistically in the Parent universe. Chapter 11 defines a model for the 23 universes of the Cosmos.

THE COSMOS THEORY SPACES SPECTRUM

HyperCosmos SECTOR SPACES SPECTRUM

Blaha Space Number $N = o_s$	Cayley-Dickson Number n	Cayley Number d_c	Dimension Array column length $d_{cN} = d_{cr}$	Dimension Array Size $d_{dN} \equiv d_{dr}$	Space-time-Dimension r	CASe Group $su(2^{r/2},2^{r/2})$ CASe	Fermion Spin s	Unification Space r'	d_{dN}'
0	10	1024	2048	2048^2	18	su(512,512)	255/2	40	2048^4
1	9	512	1024	1024^2	16	su(256,256)	127/2	36	1024^4
2	8	256	512	512^2	14	su(128,128)	63/2	32	512^4
3	7	128	256	256^2	12	su(64,64)	31/2	26	256^4
4	6	64	128	128^2	10	su(32,32)	15/2	22	128^4
5	5	32	64	64^2	8	su(16,16)	7/2	18	64^4
6	4	16	32	32^2	6	su(8,8)	3/2	14	32^4
7	**3**	**8**	**16**	$\mathbf{16^2}$	**4**	**su(4,4)**	½	**12**	$\mathbf{16^4}$
8	2	4	8	8^2	2	su(2,2)	0	8	8^4
9	1	2	4	4^2	0	su(1,1)	- ¼	4	4^4

Limos SECTOR SPACES SPECTRUM

	N	n	d_c	d_{cN}	d_{dN}	r	CASe	s	r'	d_{dN}'
Limos:	10	0	1	2	2^2	-2	U(1)	-3/8	0	0
	11	-1	½	1	1^2	-4	U(½)	-7/16	-4	0
	12	-2	¼	½	$(½)^2$	-6	U(¼)	-15/32	-8	0
	13	-3	1/8	¼	$(¼)^2$	-8	U(1/8)	- 31/64	-12	0
	14	-4	1/16	1/8	$(1/8)^2$	-10	U(1/16)	-63/128	-16	0
					⋮					
	∞	$-\infty$	0	0	0	$-\infty$	U(0)	0	$-\infty$	0

Figure 10.1. The Cosmos Theory space spectrum. Spaces with non-negative space-times may have universe instances. Spaces with negative space-times form the Limos set of Limos spaces, which are displayed in the shaded area.

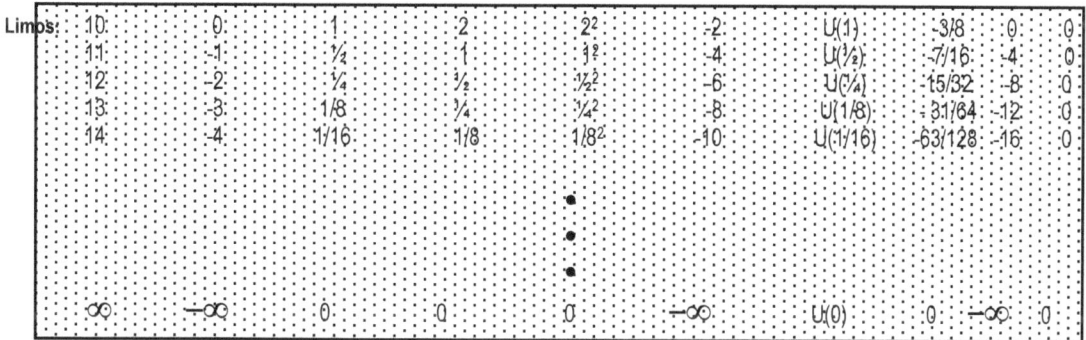

11. Planckian Distribution of Cosmos Universes

The Cosmos has ten spaces. Each space may have zero or more created universes.[79] The set of universes of the ten spaces have a distribution that we will assume is a Planckian distribution similar in a form similar to that of chapter 4. There are several questions involving the Cosmos Theory that may be addressed by considering the Planckian distribution: Why are there only ten spaces? What determines the set of universes of the ten spaces? How is the distribution of universes determined? Is the distribution statistical in nature or is there a driving force for the creation of universes of the various Cosmos spaces?

We will assume the set of Cosmos spaces are characterized by an energy determined by the ProtoCosmos Model.[80] The model is based on a scaling potential that gives an energy spectrum:

$$E_n \cong m(1 - d_{dn}) \tag{11.1}$$

where d_{dn} is the nth dimension array of Cosmos Theory

$$d_{dn} = 2^{2n+2} \tag{11.2}$$

with n being a Cayley-Dickson number. (Fig. 10.1.) We take the d_{dn} values to be "energies" of the Cosmos spaces. We parameterize the gambols of universes with the index n. We identify gambol energies $\varepsilon(s)$ in a similar manner to eq. 4.21:

$$\varepsilon(s) = (M_g s + M)/(s + 1) \tag{11.3}$$

where s is the fractionation, M is the total mass-energy[81] of our universe, and M_g = M/1024 is the gambol mass in analogy with chapter 4. The space-time dimension of each Cosmos space is

$$r = 2n - 2 \tag{11.4}$$

The ten spaces were selected by r values: The space with r = 0 specifies the smallest space. The n = 10 space where r = 18 space-time dimensions is the largest of the ten spaces.

11.1 The States of Cosmos Universes

We now develop a model for the set of Cosmos universes. We separate the set of all Cosmos universes of all spaces into a set of states characterized by their energy $\varepsilon(n)$.

---

[79] A universe is an entity with a mass-energy distribution. A universe may be defined for any of the ten spaces.
[80] See Blaha (2022f).
[81] We chose this *ad hoc* value arbitrarily. A different choice of M would lead to similar results.

The universes can transition between ProtoCosmos states through quantum jumps just as electrons in hydrogen atoms. The possible processes are:

- with probability S_{12} for a transition from E_2 to E_1 with a quantum universe (gambol) emission;
- with probability $A_{21}u$ absorption of a quantum giving a jump from E_1 to E_2 where u is the energy density;
- with probability $I_{12}u$ for induced emission of a quantum giving a jump from E_1 to E_2 where u is the energy density

Section 4.3.1 shows how these processes lead to the Planckian distribution of universe gambols:[82]

$$E_{\varepsilon n} = U_g(\varepsilon(s)) = 15\, N\, (\pi kT)^{-4}\, \varepsilon^3/(e^{\varepsilon/kT} - 1) \tag{11.5}$$

In making the above assumptions we note that we implicitly assume that transitions are possible between the universes of different spaces through inter-universe quantum interactions. We further implicitly assume that there is a Parent universe time during which a Planckian distribution of universes can evolve. It has a clock for the onset of equilibrium and for the transformations of universes subsequently

We take the universe gambol temperature T_g to be

$$kT_g(s) = 0.0785\, M\, (s + 1) \tag{11.6}$$

as in eq. 4.25 – a similar relation.

We set the fractionation number s that specifies the mass fractionation by a factor of 1/s with

$$s = 1/(an) \tag{11.7}$$

in analogy with eq. 4.22 where n is an integer (the Cayley-Dickson number of Fig. 10.12) and a is a constant.

The constant a is set to

$$a = bt_{NOW} = 0.377 \tag{11.7a}$$

where $t_{NOW} = 4.35 \times 10^{17}$ sec and $b = 8.667 \times 10^{-19}$ sec^{-1}. If all gambol universes were created at the same Parent universe time, a possibility that is not ruled out, then a may have a universal significance.

---

[82] One can view the set of all universes as composing a "black body" that may be viewed as the primary source of transitions.

Now

$$U_g(\varepsilon(s)) = U_g(\varepsilon(n)) \tag{11.8}$$

and similarly for $kT_g(s)$ and $\varepsilon(s)$. We set the values

$M = 1.712 \times 10^{81}$ GeV/c^2 - The total mass-energy of our universe (11.9)
$M_g = M/1024$
$a = 0.377 \approx \frac{1}{4}\pi^{\frac{1}{4}} = 0.333$
$N = 6.277 \times 10^{81}$ GeV/c^2 = 3.67 M

11.2 The Planckian Distribution of Universes

We can now graph the distribution of universes vs. n. We make n a continuous variable. We give U_g ($\varepsilon(n)$) the role of specifying the number of universe gambols of type n by setting the N normalization factor in eq. 11.5 appropriately.

We implement a limit on the number of n values n_0 by requiring

$$E_{\varepsilon n_0} = U_g(\varepsilon(n_0)) = 1 \tag{11.10}$$

for $n = n_0$. If $n_0 = 10$ we have a ten space Cosmos Theory since no universe gambols would exist above n_0.

We set

$$n = [U_g(\varepsilon(s))] \tag{11.11}$$

where [] indicates greatest integer less than $U_g(\varepsilon(s))$.

The $U_g(\varepsilon(n))$ distribution gives a unifying probabilistic interpretation of the set of universes. Our universe is an n = 3 gambol universe.

11.3 Plot of $U_g(\varepsilon(s))$

The universe gambol distribution is analogous to the photon plot of the average aggregate photon numbers vs. photon energies hν.

We have set the value of N to limit the number of spaces with universes to *ten* – the Cosmos set of ten spaces:

The plot of $U_g(\varepsilon(n))$ appears in Fig 11.1. The numbers of universes is plotted in Fig. 11.2. The following points are evident:

1. The plot shows a declining number of universes as n increases.

2. The value for n = 10 is the maximum universe of the Cosmos Theory spaces in this model. (Fig. 10.1). This value of n = 10 is chosen by setting N appropriately (eq. 11.9)

3. The set of $[U_g(\varepsilon(s))]$ values is understandable due to increasing universe size as n increases. Thus the number of universes declines with increasing n.

4. The n = 3 number of universes is 5. Our universe is one of the 5 universes in the theory.

5. There are four Megaverses for n = 4. One of them is our Megaverse.

6. The Planckian distribution reasonably accounts for the distribution of universes of Cosmos spaces. The statistical basis of the Planckian distribution provides an understanding of the gestation of universes in the Cosmos. The ProtoCosmos Model with its energies corresponding to Cosmos Spaces may be viewed as having a distribution of corresponding universes states rather like the electrons in a multi-electron atom.

7. The Planckian distribution is based on statistics. It assumes that universes can transition between states (n values) through quantum processes as discussed in section 11.1 and earlier books.

8. It is "comforting" to know that the number of universes is *one* for n = 7 – 10. See Fig. 11.2.

9. Fig. 11.1 shows the universes of lower n values are more numerous before declining to zero at n = 0. They have a larger fractionation factor of 1/s. Correspondingly these universes have smaller dimension arrays and smaller space-time dimensions. Universes of larger n values are less numerous. They have a smaller fractionation factor of 1/s. Correspondingly, these universes have larger dimension arrays and larger space-time dimensions.

10. There are 23 universes in this model of the Cosmos.

The Planckian distribution of universes answers the question of the nature of the process that produces universes. It is probabilistic.

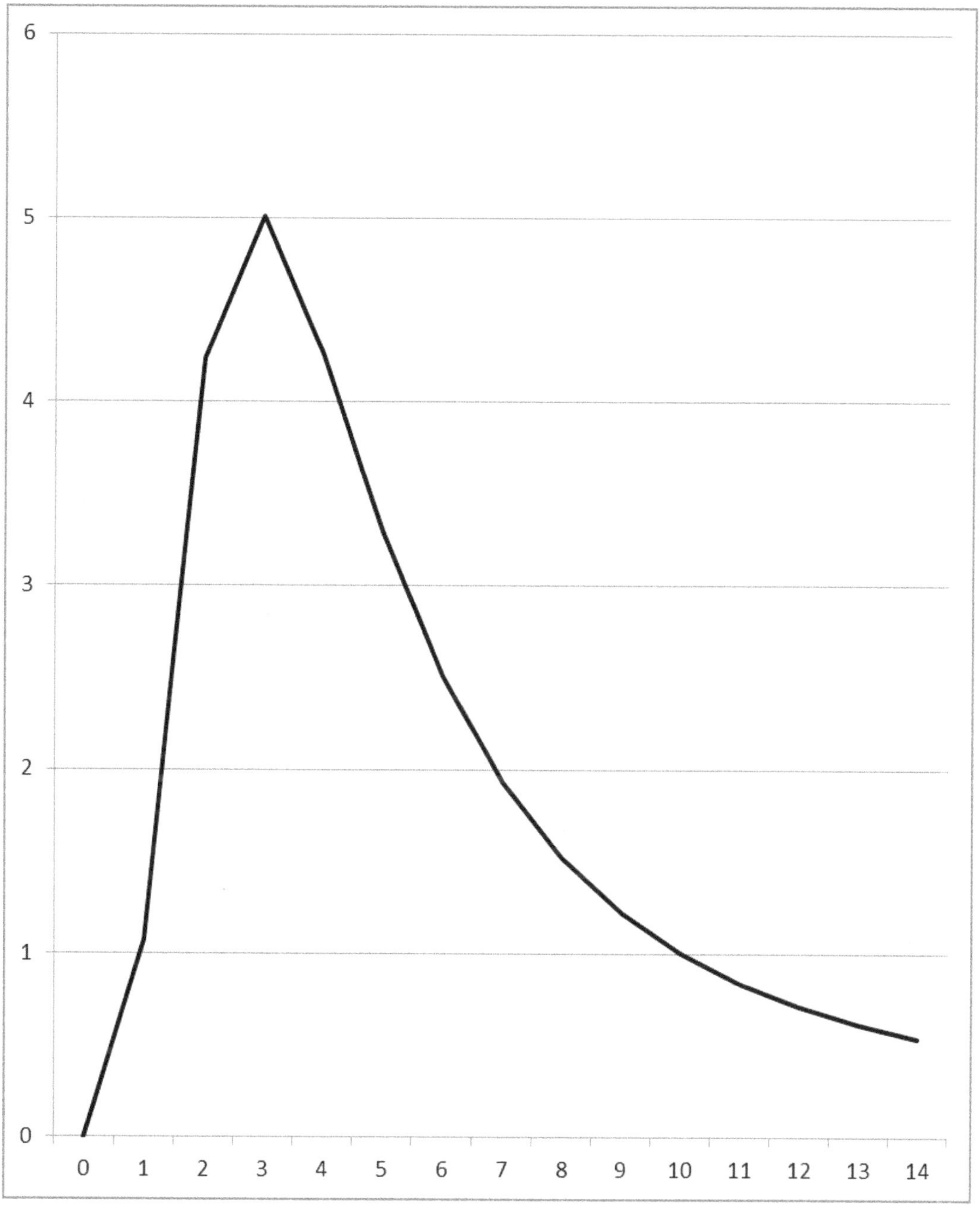

Figure 11.1. Plot of $U_g(\varepsilon(n))$ vs. the Cayley-Dickson number n. $[U_g(\varepsilon(10))] = 1$ by design. The number of universes for n > 10 is zero: $[U_g(\varepsilon(n))] = 0$ for n > 10. Small n corresponds to larger s and T_g. Large n corresponds to smaller s and T_g. The jagged curve is the result of "coarse graining" n. Fig. 11.2 shows the numbers of universes $[U_g(\varepsilon(10))]$.

UNIVERSE NUMBERS

n	1	2	3	4	5	6	7	8	9	10	11
$[E_g]$	1	4	5	4	3	2	1	1	1	1	0

Figure 11.2. Table of Cayley-Dickson number n vs. the number of universes,$[U_g(\varepsilon(n))]$. Note there are no universes for HyperCosmos spaces n > 10. Thus Cosmos Theory may be limited to ten HyperCosmos spaces with this model. The universes of type n = 1 have zero space-time dimensions and 16 components in their dimension arrays. There is no n = 0 universe. See Fig. 10.1.

12. Quark Matter Cores of Neutron Stars

The possible[83] existence of a pure quark core of enormous density within neutron stars has been raised. One suggestion is for a quark core up to a 25 km diameter weighing up to two solar masses. This core would be gravitationally bound but electromagnetically neutral due to an almost free motion of its quarks.

In this chapter we treat the quark core as an "enormous particle" that is gravitationally confined and has a gambol internal structure. It could be viewed as a black body with massive constituents (as opposed to the massless photons in a Planck black body distribution.)

The primary interaction is gravitation. Due to its effects[84] we will model the internal gambol structure based on the fractionation s being radial in nature. In the case of previous universe structure we chose s to be related to time with

$$s = 1/(bt) \tag{4.22}$$

Here we choose s to depend on the radius within the quark core. With this gambol structuring the gambol states within the quark core are a set of states[85] characterized by s initially for a discrete set of radial "levels." Then we s is made into a continuous variable.

We relate s to the radial distance r from the center of the quark core:

$$s = 1/(b'r) \tag{12.1}$$

where we chose

$$b' = 1/r_0 = 0.8 \times 10^{-6}\ \text{cm}^{-1} \tag{12.2}$$

using $r_0 = 1.25 \times 10^6$ cm as the radius of the quark core.

We set the masses with:

$$\begin{aligned} M &= \text{quark core of 2 solar masses} = 2.23 \times 10^{57}\ \text{GeV/c}^2 \\ M_g &= M/8 = 0.279 \times 10^{57}\ \text{GeV/c}^2 \end{aligned} \tag{12.3}$$

We take the energy of a radial gamble to be

$$\varepsilon(s) = (M_g\, s + M)/(s + 1) \tag{12.4}$$

---

[83] Eerneli Annala *et al*, Nature Communications (2023).
[84] The small size of the core allows us to approximate gravitational energy as constant throughout the core.
[85] Chapter 4 describes the derivation of the Planckian distribution statistically.

The average energy of all gambols of fractionation 1/s counts the number[86] *of gambols of each energy* ε. It is specified by our Planckian distribution E_ε:

$$E_\varepsilon(s) = U_g(\varepsilon(s)) = 15\, N\, (\pi kT)^{-4}\, \varepsilon^3/(e^{\varepsilon/kT} - 1) \tag{12.5}$$

where N has the dimensions GeV^2/c^4 to make E_ε have the correct dimension GeV/c^2.

We now introduce the radial parameter r using eq. 12.1 by defining ε in terms of radius r within the quark core resulting in

$$\varepsilon(r) = (M_g + b'rM)/(b'r + 1) \tag{12.6}$$

The average energy E_g for the ε energy gambols, varies with quark core radius.

We define the quark core *gambol* temperature T_{gq} with

$$kT_{gq} = 0.0785\, M(s + 1) = 0.0785\, M\, [1/(b'r) + 1] \tag{12.7}$$

Eq. 12.7 uses almost the same definition for kt as we did for particles in Blaha (2023e). The quark core gambol temperature equation is slightly modified[87] from its equation for the gambol temperature for quarks, leptons, and hadrons in Blaha (2023e):

kT_{gq} may be expressed as a function of ε using

$$b'r = [M_g - \varepsilon]/(\varepsilon - M) \tag{12.8}$$

We find

$$kT_{gq} = 0.0785\, M\, [(M - \varepsilon)/(\varepsilon - M_g) + 1] \tag{12.9}$$

Thus we can view $U_g(\varepsilon(s)) \equiv U_g(\varepsilon(t))$ as strictly a function of ε.

One expects the average energy of gambols will be very small for small r near the center of the core. Then as the radius increases the average energy of entities increases. Subsequently, near the edge of the quark core the average energy decreases.

The quark core model[88] is based on a core split by decreasing average fractionation values as r increases from zero. At each radius, viewing radial distance as discrete for the purpose of discussion, $\varepsilon(r)$ and E_g change, just as the black body Planck distribution for photons changes with photon energy hν. The energy of each gambol is

---

86 The model is analogous to the Planck black body distribution. For each photon of energy hν there are a certain number of photons. This number is specified by the black body distribution value for hν. Similarly, the average number of gambols of fractionation s is specified by E_g.

87 The insertion of 1 in this equation enables the gambol quark temperature to be a non-zero constant as $s \to 0$. The s = 0 value is outside the effective fractionation range; so we physically want $s_{effective} = 1$ as the least allowed value of s for fractionation. We extend s to zero in calculations and make it continuous to have analytic computations.

88 This somewhat sensitive calculation is performed in double precision using Excel. The calculations in Blaha (2023e) are also performed in double precision using Excel. Double precision mathematics was also required in the author's calculation of the Fine Structure Constant since it is known to thirteen decimal places. The calculation found the exact known result.

initially a 1/s part of the core energy where s is a power of 2. We then make s continuous.

Fig. 12.1 displays the average quark core energies as a function of radius.[89]

12.1 Normalization for the Quark Core Gambol Model

When s = 1 there is only one gambol. Thus it should have an energy E_g equal to the gambol energy at that radius. We chose the normalization N to make the outer edge $r_0 = 1.25 \times 10^6$ cm average energy $U_g(\varepsilon(r_0))$ equal to the gambol energy $\varepsilon(s = 1)$ at the boundary of the quark core: [90]

$$E_\varepsilon(1)/\varepsilon(1) = 1$$

We find

$$N = 2.155 \times 10^{114}\ \text{GeV}^2/c^4 = 0.434\ M^2 \qquad (12.10)$$

≈

Note: $N \approx M^2/2\ \text{GeV}^2/c^4$ is a good approximation to N. It has the correct dimensions agreeing with E_ε's dimension GeV/c^2.

The Quark Core Gambol Model (Fig. 12.1) provides a statistical fit to the core's gambol energy distribution.

[89] This value was used in Blaha (2023e).

[90] In this case s = 1 and the factorization thus gives a gambol equivalent to the entire quark core.

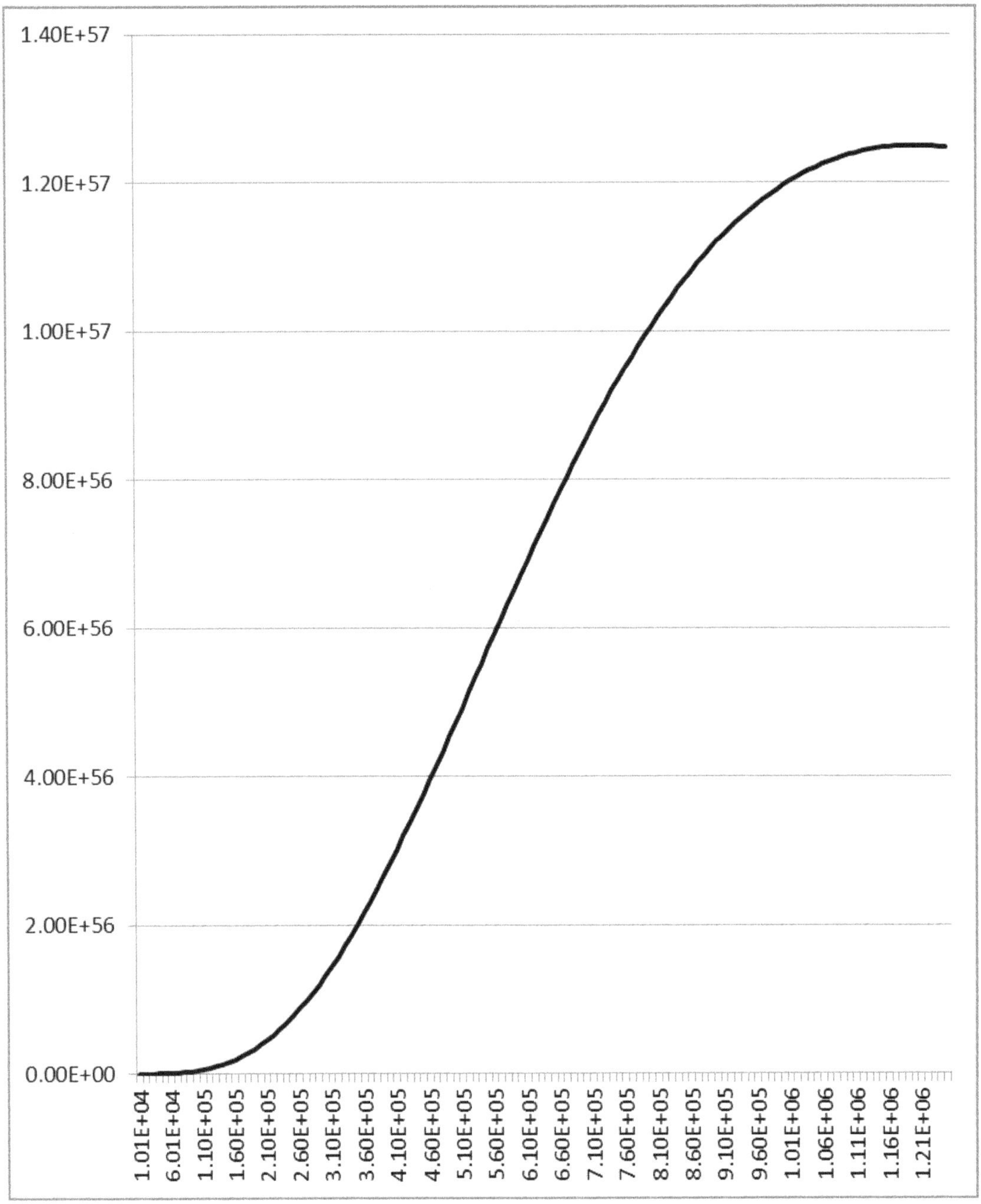

Figure 12.1. Plot of E_g quark core energy as a function of radial distance r.

13. Interpretation of b, b' and a Constants

The constants b and b' may be interpreted as specifying the physical bounds of entities. The constant b' used in chapter 12 is determined by the radial limit r_0 of the quark core. The constant b of chapters 4 and 6 was used in an analogous way to set the scale of time and thereby to determine a life time for the universe.

The analysis in chapters 4 and 6 leads us to entertain the possibility that b "sets" a limit on the lifetime of evolution of our universe. Thus

$$t_{end} = b^{-1} = 1.15 \times 10^{18} \text{ sec} \tag{13.1}$$

Since

$$t_{NOW} = 4.35 \times 10^{17} \text{ sec}$$

we find

$$t_{end} = 3.79\, t_{NOW} \equiv 52 \text{ billion years} \tag{13.2}$$

which is a long time. At t_{end} the evolution of the universe and its contents would presumably end. The models would then suggest the entry of the universe into a stable, non-expanding state—similar to that of a fundamental particle. Then a universe may be viewed as a particle in the Megaverse as we have described in the past.[91]

The constant a given by

$$a = bt_{NOW} \tag{13.3}$$

leads to the collapse of s into a Cosmos space Cayley-Dickson numbers fractionation parameterization.

[91] See Blaha (2021d).

14. Boson Gambols

There is good reason to believe bosons may also be decomposable into gambols. In this chapter we will consider the photon – ρ meson[92] transition that appears in Generalized Vector Meson Dominance (GVDM). We will consider a gsmbol model equivalent of the Higgs Mechanism that gives mass to intermediate vector bosons in ElectroWeak Theory.

Then we describe a boson particle – gambol quantum field theory. This theory describes the inheritance of parent boson features by a boson gambol. Boson theories of this kind may be used in the development of bosonic universes models corresponding to the fermion models considered earlier in this book.

14.1 Photon – ρ Gambol Transitions

In GVDM photons may be viewed as a combination of a pure electromagnetic photon and hadronic mesons. We will describe the transition between pure electromagnetic photons and ρ mesons using a gambol model similar to those of Blaha (2023e). The photon energy is E = hν and the ρ meson mass is m_ρ. We choose the effective photon gambol energy to be E/8, and set the ρ gambol mass equal to $m_\rho/8$. *We picture physical photons as having a gambol substructure.*

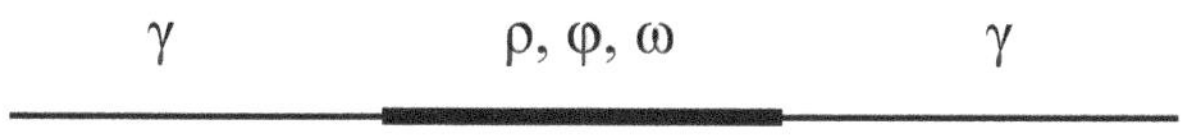

Figure. Couplings of the physical photon.

The photon-ρ transition parameters and equations are:

$E = h\nu = m_\gamma$	for the photon
$m_{g\gamma} = m_\gamma/8$	for the photon
$\varepsilon_\gamma(s) = (m_{g\gamma}\, s + m_\gamma)/(s + 1)$	for the photon

$kT_g = 0.0785\, m_\rho$
$U_g(\varepsilon(s)) = 15\, N\, (\pi kT)^{-4}\, \varepsilon^3/(e^{\varepsilon/kT} - 1)$

$\varepsilon_\rho(s) = [(m_{g\rho}\, s + m_\rho)/(s + 1)]$	for the ρ
$m_\rho = 770\ \text{mev}/c^2 = 0.77\ \text{GeV}/c^2$	for the ρ meson

[92] The dressed photon is a superposition of the pure electromagnetic photon and vector mesons. We treat the dressed photon as generating the vector meson parts through quantum transitions with certain transition probabilities. The dressed photon then interacts either purely or through a vector meson(s). The gambol distributions are estimates of the probabilities of each dressed meson part. They coincide at the vector meson mass. (See Fig. 14.1.) Similar considerations apply to ω and φ vector mesons.

$m_{g\rho} = m_\rho/8 = 0.096\ \text{GeV}/c^2$ ρ gambol mass
$N = 1$

where s is the fractionation parameter, $m_{g\rho}$ is the ρ gambol mass, m_ρ is the ρ mass, and N is a normalization factor.

Figs. 14.1 – 14.3 contain the Planckian distributions for the photon and ρ meson. Similar results hold for ω and φ distributions.

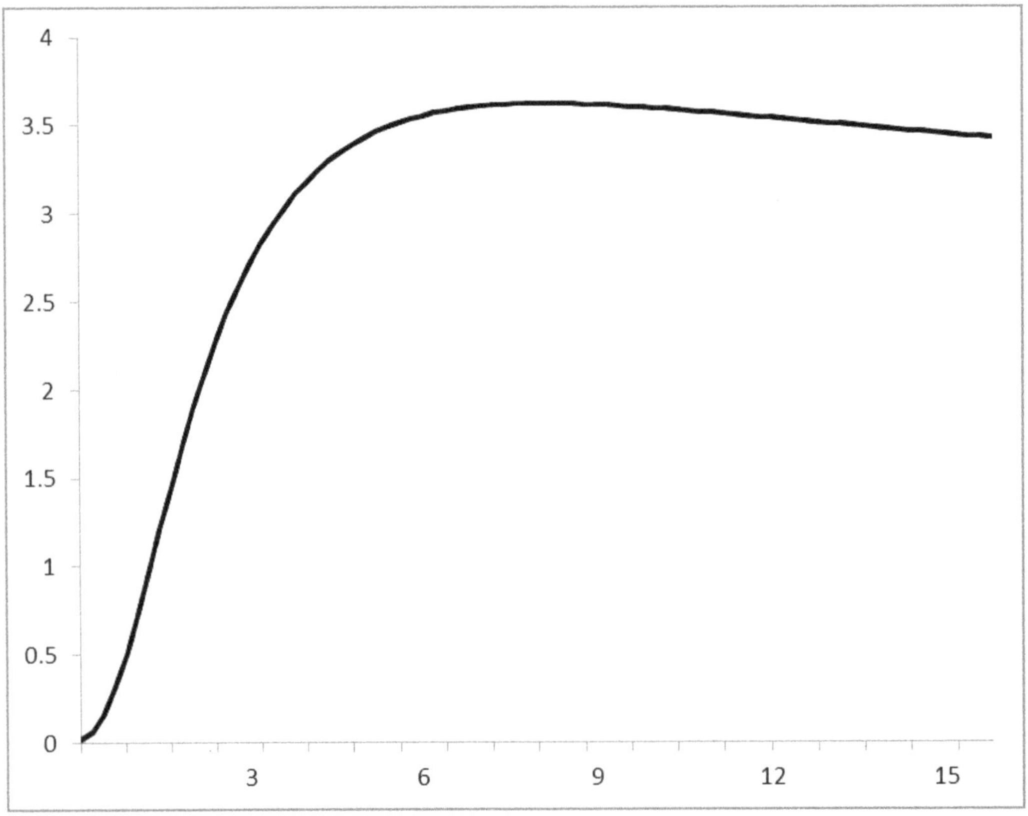

Figure 14.1. Plots of photon and ρ distributions vs. s for E = 0.77 GeV/c². Their plots are the same.

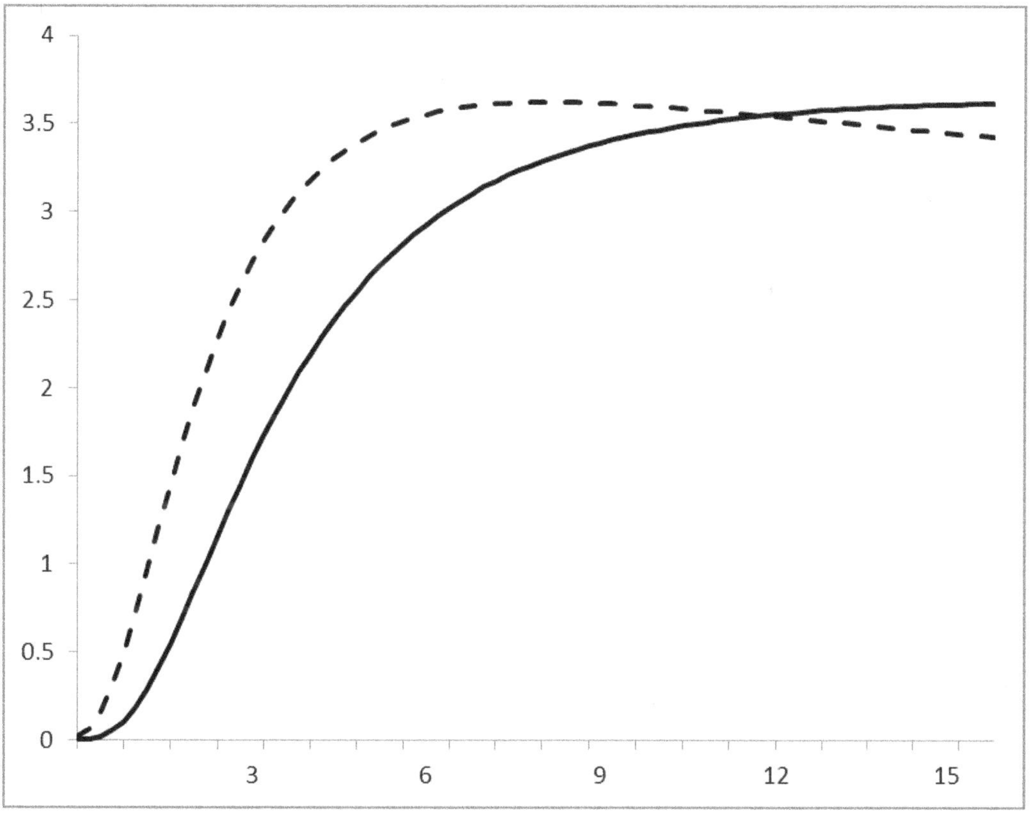

Figure 14.2. Plots of photon (solid line) and ρ (dashed line) distributions vs. s for E = 1.0 GeV/c^2.

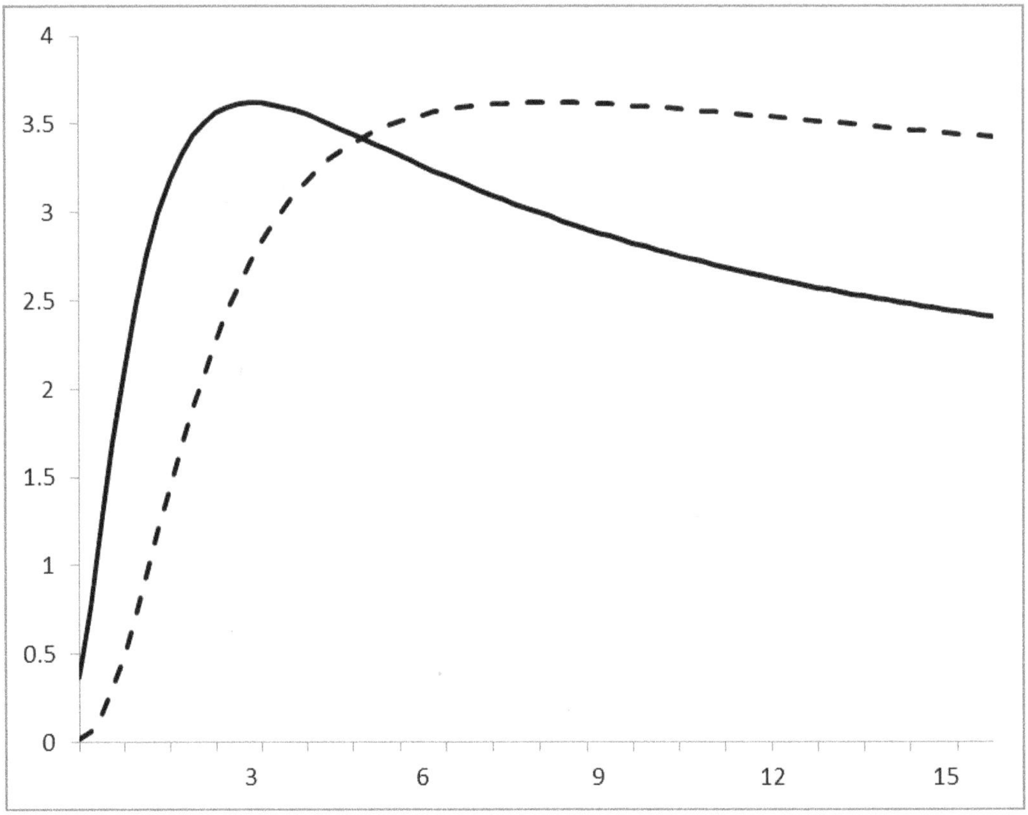

Figure 14.3. Plots of photon (solid line) and ρ (dashed line) distributions vs. s for E = 0.5 GeV/c^2.

14.2 $W^{\pm}$ Vector Boson Gambols

The $W^{\pm}$ gambol model parameters for the transition from a massless state to a massive state are:[93]

$E_{W_0} = h\nu_{W_0} = m_{W_0}$	for the *massless* $W_0^{\pm}$ vector boson energy
$m_{gW_0} = E_{W_0}/8$	*massless* $W_0^{\pm}$ gambol energy
$\varepsilon_{W_0}(s) = [(m_{gW_0}\, s + m_{W_0})/(s + 1)]$	gambol energy quantum for the *massless* $W_0^{\pm}$ vector boson

$kT_g = 0.0785\ m_W$

$U_g(\varepsilon(s)) = 15\ N\ (\pi kT)^{-4}\ \varepsilon^3/(e^{\varepsilon/kT} - 1)$

$m_W = 80.377\ \text{GeV}/c^2$	for the massive $W^{\pm}$ vector boson
$m_{gW} = m_W/8 = 10.05\ \text{GeV}/c^2$	$W^{\pm}$ gambol mass
$\varepsilon_W(s) = (m_{gW}\, s + m_W)/(s + 1)$	massive $W^{\pm}$ vector boson gambol energy quantum

$N = 1$

where s is the fractionation parameter, m_{W_0} is the effective mass of the $W_0^{\pm}$, m_{gW} is the $W^{\pm}$ gambol mass, m_W is the $W^{\pm}$ mass and N is a normalization factor.

Figs. 14.4 – 14.6 contain the Planckian distributions for the massless and massive $W^{\pm}$ cases. Similar results hold for Z^0 case.

93 One imagines adiabatically turning on the Higgs Mechanism for mass generation for each vector boson. The gambol model parallels the Higgs Mechanism process. The Z^0 gambol model is analogous.

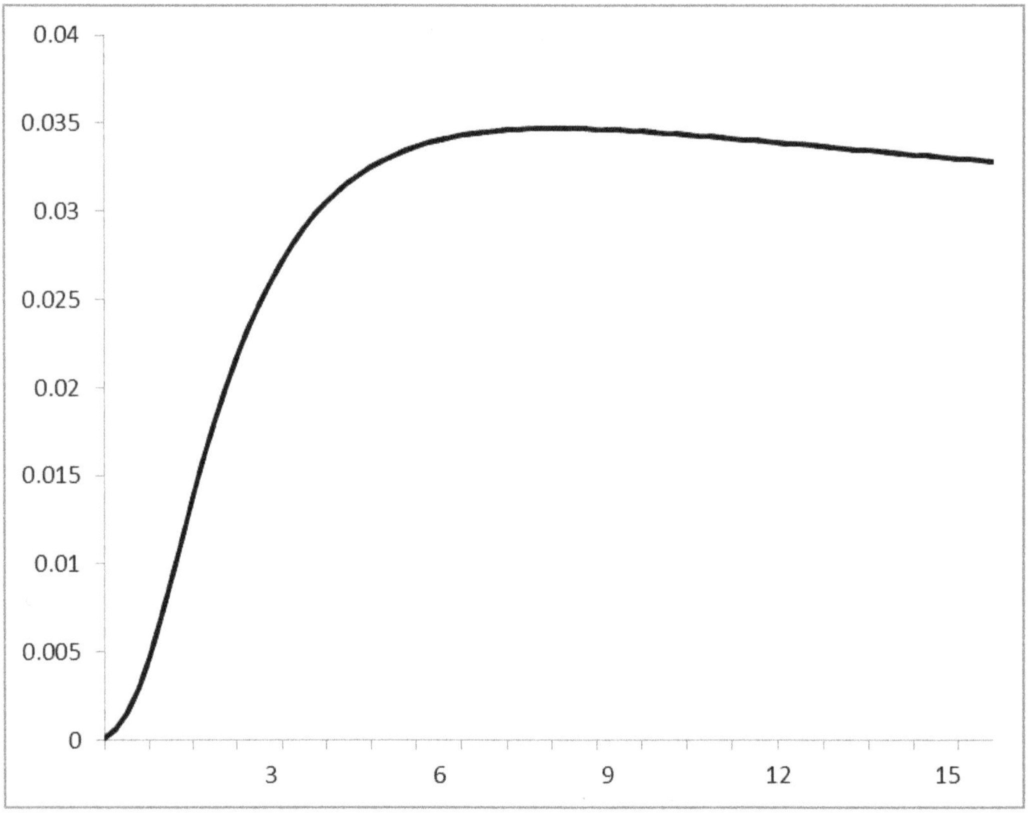

Figure 14.4. Plots of massless $W_0^{\pm}$ and massive $W^{\pm}$ distributions vs. s for E = 80.377 GeV/c^2. Their plots are the same.

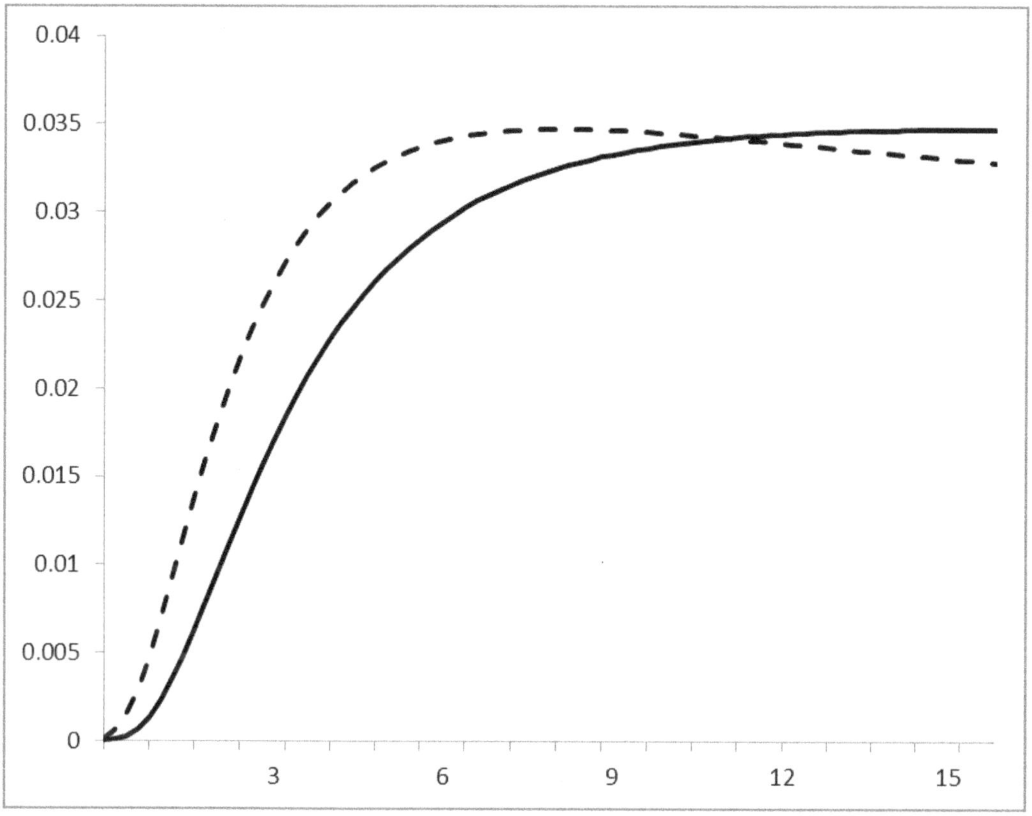

Figure 14.5. Plots of massless $W_0^{\pm}$ (solid line) and massive $W^{\pm}$ (dashed line) distributions vs. s for E = 100 GeV/c^2.

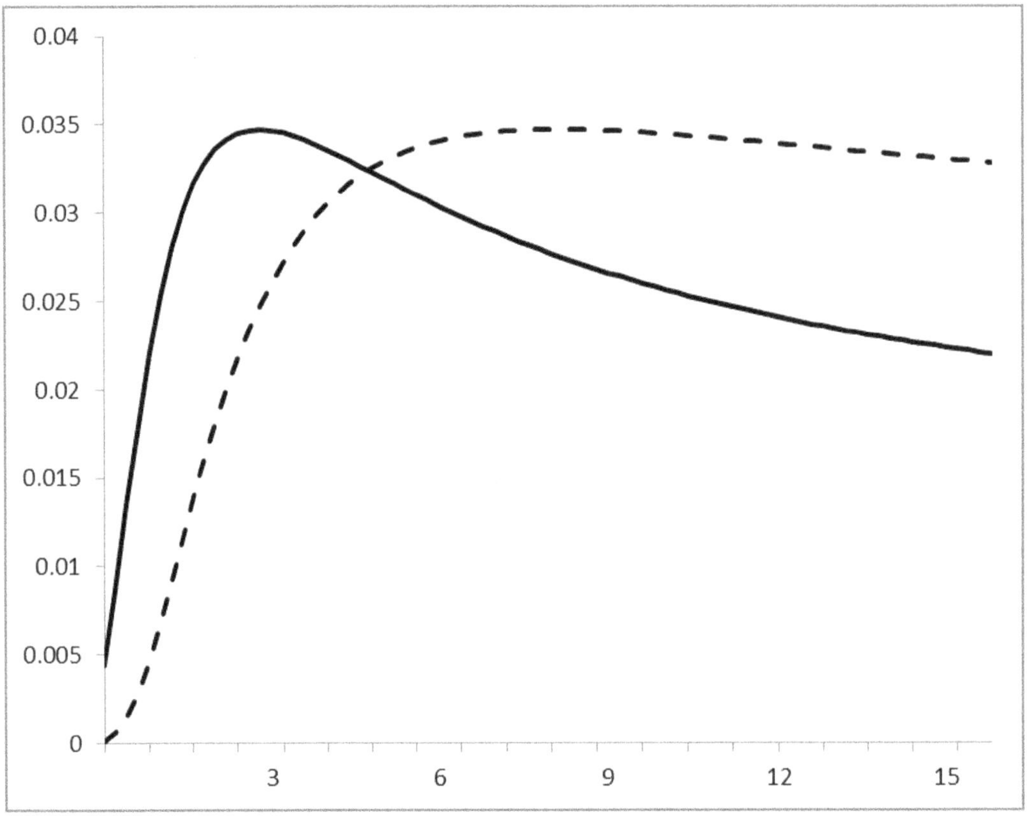

Figure 14.6. Plots of massless $W_0^{\pm}$ (solid line) and massive $W^{\pm}$ (dashed line) distributions vs. s for E = 50 GeV/c^2.

14.3 Relation to Higgs Mechanism

In sections 9.10 and 9.11 of Blaha (2023e) we found that the gambol structures within decaying particles (both the particle and its decay product) combine to approximate a Breit-Wigner type of distribution when their product[94] is plotted as a function of energy. It has a peak at the energy of the resonance. At this energy the probability distribution of the product constructively creates the peak. At nearby energies the product displays a form of destructive interference. See Figs. 9.20 – 9.22 of Blaha (2023e).

14.3.1 Photon – ρ Relation in GVDM

In the cases of the photon – ρ transition the distribution curves in Figs. 14.1 – 14.3, when multiplied figure by figure, also show constructive (Fig. 14.1) and destructive (Figs. 14.2 – 14.3) "interference." The gambol model thus simulates the relation between the photon and ρ in interactions in the GVDM models. The effective transition between the electromagnetic photon and the ρ may be determined as a function of photon energy E = hν.

14.3.2 Massless $W^{\pm}$ (and Z^0) Transition to Massive $W^{\pm}$ Vector Bosons and Higgs Mechanism

In the cases of the massless $W^{\pm}$ (and Z^0) transition to massive $W^{\pm}$ vector bosons a similar form of constructive and destructive interference appears for the distributions when they are multiplied in each of the plots of Figs. 14.4 – 14.6.

This gambol model thus simulates the transition usually implemented by the Higgs Mechanism. It shows a form of internal reorganization of the gambol substructure of the vector bosons. The effective transition between a massless vector boson and the massive vector boson may be determined as a function of massless vector boson energy $E_{W0} = h\nu_{W0}$.

14.4 Boson Particle – Gambol Lagrangian

Just as we showed that quantum field theories exist for fermions in earlier chapters we can develop equivalent boson quantum field theories following an analogous procedure.

For example the development in chapter 3 may be paralleled by defining[95] two coordinates boson fields with the form

$$\varphi_i(y, z) \tag{14.1}$$

where y and z are independent coordinates in r space-time dimensions, and where i = 1, 2 labels the scalar PseudoQuantum[96] fields.

---

[94] The product is implied by the nature of transition probabilities for S-matrix elements.

[95] We follow the conventions of Bjorken (1965) with the $g^{\mu\nu}$ metric (1, -1, -1, -1).

[96] Appendix C of Blaha (2016f) presents a first quantized PseudoQuantum theory CQ Mechanics that embodies both classical and quantum theory. **This theory is the non-relativistic quantum mechanics limit of the relativistic PseudoFermion theory.** CQ Mechanics has two sets of coordinates that combine to create a generalization of conventional Quantum Mechanics. It has applications in a generalized Feynman path integral formalism, a generalized Schrödinger equation, a generalized Boltzmann equation, the Fokker-Planck equation, a generalized approach to quantum and classical chaos, and to quantum entanglement as well as semi-quantum entanglement. Our

One begins by defining a free boson PseudoQuantum Lagrangian (scalar boson case) with two PseudoQuantum wave functions φ_1 and φ_2 that are functions of two sets of coordinates in r space-time dimensions, y and z,

$$\mathscr{L} = \partial/\partial y^\mu \partial/\partial z^\nu \varphi_1]\ \partial/\partial y_\mu \partial/\partial z_\nu \varphi_2 - m^2 \varphi_1\ \varphi_2 \tag{14.2}$$

If we introduce fractionation s where the particle is fractionated to s gambols, and internal symmetry index γ, the free boson wave function has the form:

$$\varphi_\gamma(y, z) = \sum_s \int dp^{r-1} \int dq^{r-1}\ N(p, q)[a_{\gamma i}(s,p,q)\exp(-ip\cdot y - iq\cdot z) + a_{\gamma i}\ (s,p,q)^\dagger \exp(ip\cdot y + iq\cdot z)] \tag{14.3}$$

for i = 1, 2 where N(p, q) is a normalization factor.

The composite creation and annihilation operators satisfy the anti-commutation relations:

$$[a_{\gamma i}(s, p, q), a_{\gamma j}(s', p', q')^\dagger] = (1 - \delta_{ij})\ \delta_{s,s'}\ \delta^{r-1}(\mathbf{p} - \mathbf{p'})\delta^{r-1}(\mathbf{q} - \mathbf{q'})\ U_g(\varepsilon(s,q)) \tag{14.4}$$

where U_g is given by eqs. 2. 22.

The other commutation operators are zeroes.

$$\begin{aligned}[][a_{\gamma i}(s, p, q), a_{\gamma j}(s', p', q')] &= 0 \\ [a_{\gamma i}(s, p, q)^\dagger, a_{\gamma j}(s', p', q')^\dagger] &= 0\end{aligned} \tag{14.5}$$

Extracted boson gambol and particle operators, and so on then follows analogously to that of chapter 3.

Thus a boson particle state is a superposition of boson gambol states. We then have a quantum field theory for a boson particle – gambol configuration.

"Pseudo" formalisms apply to both Quantum Field Theory and Quantum Mechanics. In these applications there is a clear almost continuous transition between the quantum and the classical sectors.

REFERENCES

Akhiezer, N. I., Frink, A. H. (tr), 1962, *The Calculus of Variations* (Blaisdell Publishing, New York, 1962).

Bjorken, J. D., Drell, S. D., 1964, *Relativistic Quantum Mechanics* (McGraw-Hill, New York, 1965).

Bjorken, J. D., Drell, S. D., 1965, *Relativistic Quantum Fields* (McGraw-Hill, New York, 1965).

Blaha, S., 1995, *C++ for Professional Programming* (International Thomson Publishing, Boston, 1995).

_______, 1998, *Cosmos and Consciousness* (Pingree-Hill Publishing, Auburn, NH, 1998 and 2002).

_______, 2002, *A Finite Unified Quantum Field Theory of the Elementary Particle Standard Model and Quantum Gravity Based on New Quantum Dimensions™ & a New Paradigm in the Calculus of Variations* (Pingree-Hill Publishing, Auburn, NH, 2002).

_______, 2004, *Quantum Big Bang Cosmology: Complex Space-time General Relativity, Quantum Coordinates™ Dodecahedral Universe, Inflation, and New Spin 0, ½, 1 & 2 Tachyons & Imagyons* (Pingree-Hill Publishing, Auburn, NH, 2004).

______, *2005a, Quantum Theory of the Third Kind: A New Type of Divergence-free Quantum Field Theory Supporting a Unified Standard Model of Elementary Particles and Quantum Gravity based on a New Method in the Calculus of Variations* (Pingree-Hill Publishing, Auburn, NH, 2005).

______, 2005b, *The Metatheory of Physics Theories, and the Theory of Everything as a Quantum Computer Language* (Pingree-Hill Publishing, Auburn, NH, 2005).

______, 2005c, *The Equivalence of Elementary Particle Theories and Computer Languages: Quantum Computers, Turing Machines, Standard Model, Superstring Theory, and a Proof that Gödel's Theorem Implies Nature Must Be Quantum* (Pingree-Hill Publishing, Auburn, NH, 2005).

______, 2006a, *The Foundation of the Forces of Nature* (Pingree-Hill Publishing, Auburn, NH, 2006).

______, 2006b, *A Derivation of ElectroWeak Theory based on an Extension of Special Relativity; Black Hole Tachyons; & Tachyons of Any Spin.* (Pingree-Hill Publishing, Auburn, NH, 2006).

_______, 2007a, *Physics Beyond the Light Barrier: The Source of Parity Violation, Tachyons, and A Derivation of Standard Model Features* (Pingree-Hill Publishing, Auburn, NH, 2007).

______, 2007b, *The Origin of the Standard Model: The Genesis of Four Quark and Lepton Species, Parity Violation, the ElectroWeak Sector, Color SU(3), Three Visible Generations of Fermions, and One Generation of Dark Matter with Dark Energy* (Pingree-Hill Publishing, Auburn, NH, 2007).

______, *2008a, A Direct Derivation of the Form of the Standard Model From GL(16) (Pingree-Hill Publishing, Auburn, NH, 2008).*

______, 2008b, *A Complete Derivation of the Form of the Standard Model With a New Method to Generate Particle Masses Second Edition* (Pingree-Hill Publishing, Auburn, NH, 2008)

______, 2009, *The Algebra of Thought & Reality: The Mathematical Basis for Plato's Theory of Ideas, and Reality Extended to Include A Priori Observers and Space-Time Second Edition* (Pingree-Hill Publishing, Auburn, NH, 2009).

______, 2010a, *Operator Metaphysics: A New Metaphysics Based on a New Operator Logic and a New Quantum Operator Logic that Lead to a Mathematical Basis for Plato's Theory of Ideas and Reality* (Pingree-Hill Publishing, Auburn, NH, 2010).

______, 2010b, *The Standard Model's Form Derived from Operator Logic, Superluminal Transformations and GL(16)* (Pingree-Hill Publishing, Auburn, NH, 2010).

______, 2010c, *SuperCivilizations: Civilizations as Superorganisms* (McMann-Fisher Publishing, Auburn, NH, 2010).

______, 2011a, *21st Century Natural Philosophy Of Ultimate Physical Reality* (McMann-Fisher Publishing, Auburn, NH, 2011).

______, 2011b, *All the Universe! Faster Than Light Tachyon Quark Starships & Particle Accelerators with the LHC as a Prototype Starship Drive Scientific Edition* (Pingree-Hill Publishing, Auburn, NH, 2011).

______, 2011c, *From Asynchronous Logic to The Standard Model to Superflight to the Stars* (Blaha Research, Auburn, NH, 2011).

______, 2012a, *From Asynchronous Logic to The Standard Model to Superflight to the Stars volume 2: Superluminal CP and CPT, U(4) Complex General Relativity and The Standard Model, Complex Vierbein General Relativity, Kinetic Theory, Thermodynamics* (Blaha Research, Auburn, NH, 2012).

______, 2012b, *Standard Model Symmetries, And Four And Sixteen Dimension Complex Relativity; The Origin Of Higgs Mass Terms* (Blaha Reasearch, Auburn, NH, 2012).

______, 2013a, *Multi-Stage Space Guns, Micro-Pulse Nuclear Rockets, and Faster-Than-Light Quark-Gluon Ion Drive Starships* (Blaha Research, Auburn, NH, 2013).

______, 2013b, *The Bridge to Dark Matter; A New Sibling Universe; Dark Energy; Inflatons; Quantum Big Bang; Superluminal Physics; An Extended Standard Model Based on Geometry* (Blaha Reasearch, Auburn, NH, 2013).

______, 2014a, *Universes and Megaverses: From a New Standard Model to a Physical Megaverse; The Big Bang; Our Sibling Universe's Wormhole; Origin of the Cosmological Constant, Spatial Asymmetry of the Universe, and its Web of Galaxies; A Baryonic Field between Universes and Particles; Megaverse Extended Wheeler-DeWitt Equation* (Blaha Reasearch, Auburn, NH, 2014).

______, 2014b, *All the Megaverse! Starships Exploring the Endless Universes of the Cosmos Using the Baryonic Force* (Blaha Research, Auburn, NH, 2014).

______, 2014c, *All the Megaverse! II Between Megaverse Universes: Quantum Entanglement Explained by the Megaverse Coherent Baryonic Radiation Devices – PHASERs Neutron Star Megaverse Slingshot Dynamics Spiritual and UFO Events, and the Megaverse Microscopic Entry into the Megaverse* (Blaha Research, Auburn, NH, 2014).

______, 2015a, *PHYSICS IS LOGIC PAINTED ON THE VOID: Origin of Bare Masses and The Standard Model in Logic, U(4) Origin of the Generations, Normal and Dark Baryonic Forces, Dark Matter, Dark Energy, The Big Bang, Complex General Relativity, A Megaverse of Universe Particles* (Blaha Research, Auburn, NH, 2015).

______, 2015b, *PHYSICS IS LOGIC Part II: The Theory of Everything, The Megaverse Theory of Everything, U(4)⊗U(4) Grand Unified Theory (GUT), Inertial Mass = Gravitational Mass, Unified Extended Standard Model and a New Complex General Relativity with Higgs Particles, Generation Group Higgs Particles* (Blaha Research, Auburn, NH, 2015).

______, 2015c, *The Origin of Higgs ("God") Particles and the Higgs Mechanism: Physics is Logic III, Beyond Higgs – A Revamped Theory With a Local Arrow of Time, The Theory of Everything Enhanced, Why Inertial Frames are Special, Universes of the Mind* (Blaha Research, Auburn, NH, 2015).

______, 2015d, *The Origin of the Eight Coupling Constants of The Theory of Everything: U(8) Grand Unified Theory of Everything (GUTE), S^8 Coupling Constant Symmetry, Space-Time Dependent Coupling Constants, Big Bang Vacuum Coupling Constants, Physics is Logic IV* (Blaha Research, Auburn, NH, 2015).

______, 2016a, *New Types of Dark Matter, Big Bang Equipartition, and A New U(4) Symmetry in the Theory of Everything: Equipartition Principle for Fermions, Matter is 83.33% Dark, Penetrating the Veil of the Big Bang, Explicit QFT Quark Confinement and Charmonium, Physics is Logic V* (Blaha Research, Auburn, NH, 2016).

______, 2016b, *The Periodic Table of the 192 Quarks and Leptons in The Theory of Everything: The U(4) Layer Group, Physics is Logic VI* (Blaha Research, Auburn, NH, 2016).

______, 2016c, *New Boson Quantum Field Theory, Dark Matter Dynamics, Dark Matter Fermion Layer Mixing, Genesis of Higgs Particles, New Layer Higgs Masses, Higgs Coupling Constants, Non-Abelian Higgs Gauge Fields, Physics is Logic VII* (Blaha Research, Auburn, NH, 2016).

______, 2016d, *Unification of the Strong Interactions and Gravitation: Quark Confinement Linked to Modified Short-Distance Gravity; Physics is Logic VIII* (Blaha Research, Auburn, NH, 2016).

______, 2016e, *MoND: Unification of the Strong Interactions and Gravitation II, Quark Confinement Linked to Large-Scale Gravity, Physics is Logic IX* (Blaha Research, Auburn, NH, 2016).

______, 2016f, *CQ Mechanics: A Unification of Quantum & Classical Mechanics, Quantum/Semi-Classical Entanglement, Quantum/Classical Path Integrals, Quantum/Classical Chaos* (Blaha Research, Auburn, NH, 2016).

______, 2016g, *GEMS Unified Gravity, ElectroMagnetic and Strong Interactions: Manifest Quark Confinement, A Solution for the Proton Spin Puzzle, Modified Gravity on the Galactic Scale* (Pingree Hill Publishing, Auburn, NH, 2016).

______, 2016h, *Unification of the Seven Boson Interactions based on the Riemann-Christoffel Curvature Tensor* (Pingree Hill Publishing, Auburn, NH, 2016).

______, 2017a, *Unification of the Eleven Boson Interactions based on 'Rotations of Interactions'* (Pingree Hill Publishing, Auburn, NH, 2017).

______, 2017b, *The Origin of Fermions and Bosons, and Their Unification* (Pingree Hill Publishing, Auburn, NH, 2017).

______, 2017c, *Megaverse: The Universe of Universes* (Pingree Hill Publishing, Auburn, NH, 2017).

______, 2017d, *SuperSymmetry and the Unified SuperStandard Model* (Pingree Hill Publishing, Auburn, NH, 2017).

______, 2017e, *From Qubits to the Unified SuperStandard Model with Embedded SuperStrings: A Derivation* (Pingree Hill Publishing, Auburn, NH, 2017).

______, 2017f, *The Unified SuperStandard Model in Our Universe and the Megaverse: Quarks, … ,* (Pingree Hill Publishing, Auburn, NH, 2017).

______, 2018a, *The Unified SuperStandard Model and the Megaverse SECOND EDITION A Deeper Theory based on a New Particle Functional Space that Explicates Quantum Entanglement Spookiness (Volume 1)* (Pingree Hill Publishing, Auburn, NH, 2018).

______, 2018b, *Cosmos Creation: The Unified SuperStandard Model, Volume 2, SECOND EDITION* (Pingree Hill Publishing, Auburn, NH, 2018).

______, 2018c, *God Theory (*Pingree Hill Publishing, Auburn, NH, 2018).

______, 2018d, *Immortal Eye: God Theory: Second Edition* (Pingree Hill Publishing, Auburn, NH, 2018).

______, 2018e, *Unification of God Theory and Unified SuperStandard Model THIRD EDITION* (Pingree Hill Publishing, Auburn, NH, 2018).

______, 2019a, *Calculation of: QED α = 1/137, and Other Coupling Constants of the Unified SuperStandard Theory* (Pingree Hill Publishing, Auburn, NH, 2019).

______, 2019b, *Coupling Constants of the Unified SuperStandard Theory SECOND EDITION* (Pingree Hill Publishing, Auburn, NH, 2019).

______, 2019c, *New Hybrid Quantum Big_Bang–Megaverse_Driven Universe with a Finite Big Bang and an Increasing Hubble Constant* (Pingree Hill Publishing, Auburn, NH, 2019).

______, 2019d, *The Universe, The Electron and The Vacuum* (Pingree Hill Publishing, Auburn, NH, 2019).

______, 2019e, *Quantum Big Bang – Quantum Vacuum Universes (Particles)* (Pingree Hill Publishing, Auburn, NH, 2019).

______, 2019f, *The Exact QED Calculation of the Fine Structure Constant Implies ALL 4D Universes have the Same Physics/Life Prospects* (Pingree Hill Publishing, Auburn, NH, 2019).

______, 2019g, *Unified SuperStandard Theory and the SuperUniverse Model: The Foundation of Science* (Pingree Hill Publishing, Auburn, NH, 2019).

______, 2020a, *Quaternion Unified SuperStandard Theory (The QUeST) and Megaverse Octonion SuperStandard Theory (MOST)* (Pingree Hill Publishing, Auburn, NH, 2020).

______, 2020b, *United Universes Quaternion Universe - Octonion Megaverse* (Pingree Hill Publishing, Auburn, NH, 2020).

______, 2020c, *Unified SuperStandard Theories for Quaternion Universes & The Octonion Megaverse* (Pingree Hill Publishing, Auburn, NH, 2020).

______, 2020d, *The Essence of Eternity: Quaternion & Octonion SuperStandard Theories* (Pingree Hill Publishing, Auburn, NH, 2020).

______, 2020e, *The Essence of Eternity II* (Pingree Hill Publishing, Auburn, NH, 2020).

______, 2020f, *A Very Conscious Universe* (Pingree Hill Publishing, Auburn, NH, 2020).

______, 2020g, *Hypercomplex Universe* (Pingree Hill Publishing, Auburn, NH, 2020).

______, 2020h, *Beneath the Quaternion Universe* (Pingree Hill Publishing, Auburn, NH, 2020).

______, 2020i, *Why is the Universe Real? From Quaternion & Octonion to Real Coordinates* (Pingree Hill Publishing, Auburn, NH, 2020).

______, 2020j, *The Origin of Universes: of Quaternion Unified SuperStandard Theory (QUeST); and of the Octonion Megaverse (UTMOST)* (Pingree Hill Publishing, Auburn, NH, 2020).

______, 2020k, *The Seven Spaces of Creation: Octonion Cosmology* (Pingree Hill Publishing, Auburn, NH, 2020).

______, 2020l, *From Octonion Cosmology to the Unified SuperStandard Theory of Particles* (Pingree Hill Publishing, Auburn, NH, 2020).

______, 2021a, *Pioneering the Cosmos* (Pingree Hill Publishing, Auburn, NH, 2021).

______, 2021b, *Pioneering the Cosmos II* (Pingree Hill Publishing, Auburn, NH, 2021).

______, 2021c, *Beyond Octonion Cosmology* (Pingree Hill Publishing, Auburn, NH, 2021).

______, 2021d, *Universes are Particles* (Pingree Hill Publishing, Auburn, NH, 2021).

______, 2021e, *Octonion-like dna-based life, Universe expansion is decay, Emerging New Physics* (Pingree Hill Publishing, Auburn, NH, 2021).

______, 2021f, *The Science of Creation New Quantum Field Theory of Spaces* (Pingree Hill Publishing, Auburn, NH, 2021).

______, 2021g, *Quantum Space Theory With Application to Octonion Cosmology & Possibly To Fermionic Condensed Matter* (Pingree Hill Publishing, Auburn, NH, 2021).
______, 2021h, *21st Century Natural Philosophy of Octonion Cosmology , and Predestination, Fate, and Free Will* (Pingree Hill Publishing, Auburn, NH, 2021).

______, 2021i, *Beyond Octonion Cosmology II : Origin of the Quantum; A New Generalized Field Theory (GiFT); A Proof of the Spectrum of Universes; Atoms in Higher Universes* (Pingree Hill Publishing, Auburn, NH, 2021).

______, 2021j, *Integration of General Relativity and Quantum Theory: Octonion Cosmology, GiFT, Creation/Annihilation Spaces CASe, Reduction of Spaces to a Few Fermions and Symmetries in Fundamental Frames* (Pingree Hill Publishing, Auburn, NH, 2021).

______, 2022a, *New View of Octonion Cosmology Based on the Unification of General Relativit and Quantum Theory* (Pingree Hill Publishing, Auburn, NH, 2022).

______, 2022b, *The Dust Beneath Hypercomplex Cosmology* (Pingree Hill Publishing, Auburn, NH, 2022).

______, 2022c, *Passing Through Nature to Eternity: ProtoCosmos, HyperCosmos, Unified SuperStandard Theory* (Pingree Hill Publishing, Auburn, NH, 2022).

______, 2022d, *HyperCosmos Fractionation and Fundamental Reference Frame Based Unification: Particle Inner Space Basis of Parton and Dual Resonance Models* (Pingree Hill Publishing, Auburn, NH, 2022).

______, 2022e, *A New UniDimension ProtoCosmos and SuperString F-Theory Relation to the HyperCosmos* (Pingree Hill Publishing, Auburn, NH, 2022).

______, 2022f, *The Cosmic Panorama: ProtoCosmos, HyperCosmos, Unified SuperStandard Theory (UST) Derivation* (Pingree Hill Publishing, Auburn, NH, 2022).

_____, 2022g, *Ultimate Origin: ProtoCosmos and HyperCosmos* (Pingree Hill Publishing, Auburn, NH, 2022).

______, 2023a, *UltraUnification and the Generation of the Cosmos* (Pingree Hill Publishing, Auburn, NH, 2023).

______, 2023b, *God and and Cosmos Theory* (Pingree Hill Publishing, Auburn, NH, 2023).

______, 2023c, *A New Completely Geometric SU(8) Cosmos Theory; New PseudoFermion Fields; Fibonacci-like Dimension Arrays; Ramsey Number Approximation* (Pingree Hill Publishing, Auburn, NH, 2023).

______, 2023d, *Newton's Apple is Now the Fermion* (Pingree Hill Publishing, Auburn, NH, 2023).

______, 2023e,*Cosmos Theory: The Sub-Particle Gambol Model* (Pingree Hill Publishing, Auburn, NH, 2023).

Eddington, A. S., 1952, *The Mathematical Theory of Relativity* (Cambridge University Press, Cambridge, U.K., 1952).

Fant, Karl M., 2005, *Logically Determined Design: Clockless System Design With NULL Convention Logic* (John Wiley and Sons, Hoboken, NJ, 2005).

Feinberg, G. and Shapiro, R., 1980, *Life Beyond Earth: The Intelligent Earthlings Guide to Life in the Universe* (William Morrow and Company, New York, 1980).

Gelfand, I. M., Fomin, S. V., Silverman, R. A. (tr), 2000, *Calculus of Variations* (Dover Publications, Mineola, NY, 2000).

Giaquinta, M., Modica, G., Souchek, J., 1998, *Cartesian Coordinates in the Calculus of Variations* Volumes I and II (Springer-Verlag, New York, 1998).

Giaquinta, M., Hildebrandt, S., 1996, *Calculus of Variations* Volumes I and II (Springer-Verlag, New York, 1996).

Gradshteyn, I. S. and Ryzhik, I. M., 1965, *Table of Integrals, Series, and Products* (Academic Press, New York, 1965).

Heitler, W., 1954, *The Quantum Theory of Radiation* (Claendon Press, Oxford, UK, 1954).

Huang, Kerson, 1992, *Quarks, Leptons & Gauge Fields 2nd Edition* (World Scientific Publishing Company, Singapore, 1992).

Jost, J., Li-Jost, X., 1998, *Calculus of Variations* (Cambridge University Press, New York, 1998).

Kaku, Michio, 1993, *Quantum Field Theory*, (Oxford University Press, New York, 1993).

Kirk, G. S. and Raven, J. E., 1962, *The Presocratic Philosophers* (Cambridge University Press, New York, 1962).

Landau, L. D. and Lifshitz, E. M., 1987, *Fluid Mechanics 2nd Edition*, (Pergamon Press, Elmsford, NY, 1987).

Misner, C. W., Thorne, K. S., and Wheeler, J. A., 1973, *Gravitation* (W. H. Freeman, New York, 1973).

Rescher, N., 1967, *The Philosophy of Leibniz* (Prentice-Hall, Englewood Cliffs, NJ, 1967).

Rieffel, Eleanor and Polak, Wolfgang, 2014, *Quantum Computing* (MIT Press, Cambridge, MA, 2014).

Riesz, Frigyes and Sz.-Nagy, Béla, 1990, *Functional Analysis* (Dover Publications, New York, 1990).

Sagan, H., 1993, *Introduction to the Calculus of Variations* (Dover Publications, Mineola, NY, 1993).

Sakurai, J. J., 1964, *Invariance Principles and Elementary Particles* (Princeton University Press, Princeton, NJ, 1964).

Weinberg, S., 1972, *Gravitation and Cosmology* (John Wiley and Sons, New York, 1972).

Weinberg, S., 1995, *The Quantum Theory of Fields Volume I* (Cambridge University Press, New York, 1995).

INDEX

About the Author

Stephen Blaha is a well-known Physicist and Man of Letters with interests in Science, Society and civilization, the Arts, and Technology. He had an Alfred P. Sloan Foundation scholarship in college. He received his Ph.D. in Physics from Rockefeller University. He has served on the faculties of several major universities. He was also a Member of the Technical Staff at Bell Laboratories, a manager at the Boston Globe Newspaper, a Director at Wang Laboratories, and President of Blaha Software Inc. and of Janus Associates Inc. (NH).

Among other achievements he was a co-discoverer of the "r potential" for heavy quark binding developing the first (and still the only demonstrable) non-Aeolian gauge theory with an "r" potential; first suggested the existence of topological structures in superfluid He-3; first proposed Yang-Mills theories would appear in condensed matter phenomena with non-scalar order parameters; first developed a grammar-based formalism for quantum computers and applied it to elementary particle theories; first developed a new form of quantum field theory without divergences (thus solving a major 60 year old problem that enabled a unified theory of the Standard Model and Quantum Gravity without divergences to be developed); first developed a formulation of complex General Relativity based on analytic continuation from real space-time; first developed a generalized non-homogeneous Robertson-Walker metric that enabled a quantum theory of the Big Bang to be developed without singularities at t = 0; first generalized Cauchy's theorem and Gauss' theorem to complex, curved multi-dimensional spaces; received Honorable Mention in the Gravity Research Foundation Essay Competition in 1978; first developed a physically acceptable theory of faster-than-light particles; first derived a composition of extremums method in the Calculus of Variations; first quantitatively suggested that inflationary periods in the history of the universe were not needed; first proved Gödel's Theorem implies Nature must be quantum; provided a new alternative to the Higgs Mechanism, and Higgs particles, to generate masses; first showed how to resolve logical paradoxes including Gödel's Undecidability Theorem by developing Operator Logic and Quantum Operator Logic; first developed a quantitative harmonic oscillator-like model of the life cycle, and interactions, of civilizations; first showed how equations describing superorganisms also apply to civilizations. A recent book shows his theory applies successfully to the past 14 years of history and to *new* archaeological data on Andean and Mayan civilizations as well as Early Anatolian and Egyptian civilizations.

He first developed an axiomatic derivation of the form of The Standard Model from geometry – space-time properties – The Unified SuperStandard Model. It unifies all the known forces of Nature. It also has a Dark Matter sector that includes a Dark ElectroWeak sector with Dark doublets and Dark gauge interactions. It uses quantum coordinates to remove infinities that crop up in most interacting quantum field theories and additionally to remove the infinities that appear in the Big Bang and generate inflationary growth of the universe. It shows gravity has a MOND-like form without sacrificing Newton's Laws. It relates the interactions of the MOND-like sector of gravity with the r-potential of Quark Confinement. The axioms of the theory lead to the question of their origin. We suggest in the preceding edition of this book it can be attributed to an entity with God-like properties. We explore these properties in "God Theory" and show they predict that the Cosmos exists forever although individual universes (or incarnations of our universe) "come and go." Several other important results emerge from God Theory such a functionally triune God. The Unified SuperStandard Theory has many other important parts described in the Current Edition of *The Unified SuperStandard Theory* and expanded in subsequent volumes.

Blaha has had a major impact on a succession of elementary particle theories: his Ph.D. thesis (1970), and papers, showed that quantum field theory calculations to all orders in ladder approximations could not give scaling deep inelastic electron-nucleon scattering. He later showed the eigenvalue equation for the fine structure constant α in Johnson-Baker-Willey QED had a zero at $\alpha = 1$ not 1/137 by solving the Schwinger-Dyson equations to all orders in an approximation that agreed with exact results to 4th order in α thus ending interest in this theory. In 1979 at Prof. Ken Johnson's (MIT) suggestion he calculated the proton-neutron mass difference in the MIT bag model and found the result had the wrong sign reducing interest in the bag model. These results all appear in Physical Review papers. In the 2000's he repeatedly pointed out the shortcomings of SuperString theory and showed that The Standard Model's form could be derived from space-time geometry by an extension of Lorentz transformations to faster than light transformations. This deeper space-time basis greatly increases the possibility that it is part of THE fundamental theory. Recently, Blaha showed that the Weak interactions differed significantly from the Strong, electromagnetic and gravitation interactions in important respects while these interactions had similar features, and suggested that ElectroWeak theory, which is essentially a glued union of the Weak interactions and Electromagnetism, possibly modulo unknown Higgs particle features, be replaced by a unified theory of the other interactions combined with a stand-alone Weak interaction theory. Blaha also showed that, if Charmonium calculations are taken seriously, the Strong interaction coupling constant is only a factor of five larger than the electromagnetic coupling constant, and thus Strong interaction perturbation theory would make sense and yield physically meaningful results.

In graduate school (1965-71) he wrote substantial papers in elementary particles and group theory: The Inelastic E- P Structure Functions in a Gluon Model. Phys. Lett. B40:501-502,1972; Deep-Inelastic E-P Structure Functions In A Ladder Model With Spin 1/2 Nucleons, Phys.Rev. D3:510-523,1971; Continuum Contributions To The Pion Radius, Phys. Rev. 178:2167-2169,1969; Character Analysis of U(N) and SU(N), J. Math. Phys. 10, 2156 (1969); and The Calculation of the Irreducible Characters of the Symmetric Group in Terms of the Compound Characters, (Published as Blaha's Lemma in D. E. Knuth's book: *The Art of Computer Programming Vols. 1 – 4*).

In the early 1980's Blaha was also a pioneer in the development of UNIX for financial, scientific and Internet applications: benchmarked UNIX versions showing that block size was critical for UNIX performance, developing financial modeling software, starting database benchmarking comparison studies, developing Internet-like UNIX networking (1982) and developing a hybrid shell programming technique (1982) that was a precursor to the PERL programming language. He was also the manager of the AT&T ten-year future products development database. His work helped lead to commercial UNIX on computers such as Sun Micros, IBM AIX minis, and Apple computers.
In the 1980's he pioneered the development of PC Desktop Publishing on laser printers and was nominated for three "Awards for Technical Excellence" in 1987 by PC Magazine for PC software products that he designed and developed.
Recently he has developed a theory of Megaverses – actual universes of which our universe is one – with quantum particle-like properties based on the Wheeler-DeWitt equation of Quantum Gravity. He has developed a theory of a baryonic force, which had been conjectured many years ago, and estimated the strength of the force based on discrepancies in measurements of the gravitational constant G. This force, operative in D-dimensional space, can be used to escape from our universe in "uniships" which are the equivalent of the faster-than-light starships proposed in the author's earlier books. Thus travel to other universes, as well as to other stars is possible.
Blaha also considered the complexified Wheeler-DeWitt equation and showed that its limitation to real-valued coordinates and metrics generated a Cosmological Constant in the Einstein equations.
The author has also recently written a series of books on the serious problems of the United States and their solution as well as a book on the decline of Mankind that will follow from current social and genetic trends in Mankind.

In the past twenty years Dr. Blaha has written over 80 books on a wide range of topics. Some recent major works are: *From Asynchronous Logic to The Standard Model to Superflight to the Stars*, *All the Universe!*, *SuperCivilizations: Civilizations as Superorganisms*, *America's Future: an Islamic Surge, ISIS, al Qaeda, World Epidemics, Ukraine, Russia-China Pact, US Leadership Crisis*, *The Rises and Falls of Man – Destiny – 3000 AD: New Support for a Superorganism MACRO-THEORY of CIVILIZATIONS From CURRENT WORLD TRENDS and NEW Peruvian, Pre-Mayan, Mayan, Anatolian, and Early Egyptian Data, with a Projection to 3000 AD*, and *Mankind in Decline: Genetic Disasters, Human-Animal Hybrids, Overpopulation, Pollution, Global Warming, Food and Water Shortages, Desertification, Poverty, Rising Violence, Genocide, Epidemics, Wars, Leadership Failure.*

He has taught approximately 4,000 students in undergraduate, graduate, and postgraduate corporate education courses primarily in major universities, and large companies and government agencies.

He developed a quantum theory, The Unified SuperStandard Theory (UST), which describes elementary particles in detail without the difficulties of conventional quantum field theory. He found that the internal symmetries of this theory could be exactly derived from an octonion theory called QUeST. He further found that another octonion theory (UTMOST) describes the Megaverse. It can hold QUeST universes such as our own universe. It has an internal symmetry structure which is a superset of the QUeST internal symmetries.

Recently he developed Octonion Cosmology. He replaced it with HyperCosmos theory, which has significantly better features. He developed a fractionalization process for dimensions, particles and symmetry groups. He also described transformation that reduced particles and dimensions to a far more compact form. He also developed a precursor theory ProtoCosmos that leads to the HyperCosmos.

The author showed that space-time and Internal Symmetries can be unified in any of the ten HyperCosmos spaces in their associated HyperUnification spaces. The combined set of HyperUnification spaces enable all HyperCosmos dimensions to be obtained by a General Relativistic transformation from one primordial dimension in the 42 space-time dimension unified HyperUnification space.

At present the author devel;oped the Cosmos Theory that incorporates ProtoCosmos Theory, HyperCosmos Theory, Limos Theory, Second Kind HyperCosmos Theory and HyperUnification Spaces. He has introduced PseudoFermion wave functions and theory, He has related Cosmos Theory to Regge trajectories of spaces, parton theory, Veneziano amplitudes, Fibonacci numbers and Ramsey numbers. He has calculated an approximation to the difficult R(n,n) Ramsey numbers.

He has developed a Gambol Model that successfully accounts for e-p deep inelastic scattering, fundamental particle resonances, hadron scattering, and the inner structure of particles based on confinement through Casimir forces of ideal gambol gases. The Gambol Planckian Distribution was derived.

www.ingramcontent.com/pod-product-compliance
Lightning Source LLC
Chambersburg PA
CBHW040138200326
41458CB00025B/6300
* 9 7 8 1 7 3 5 6 7 9 5 2 5 *